MW00835321

Optical Networks

Series editor
Biswanath Mukherjee, Davis, California, USA

More information about this series at http://www.springer.com/series/6976

Hemani Kaushal • V.K. Jain • Subrat Kar

Free Space Optical Communication

Hemani Kaushal
Electronics and Communication
The NorthCap University
Gurgaon, Haryana, India

V.K. Jain
Electrical Engineering
Indian Institute of Technology Delhi
New Delhi, India

Subrat Kar
Electrical Engineering
Indian Institute of Technology Delhi
New Delhi, India

ISSN 1935-3839 ISSN 1935-3847 (electronic)
Optical Networks
ISBN 978-81-322-3689-4 ISBN 978-81-322-3691-7 (eBook)
DOI 10.1007/978-81-322-3691-7

Library of Congress Control Number: 2016963412

Printed on acid-free paper

This Springer imprint is published by Springer Nature
The registered company is Springer (India) Pvt. Ltd.
The registered company address is: 7th Floor, Vijaya Building, 17 Barakhamba Road, New Delhi 110 001, India

Thanks to our families for their affection and endless support!!

Preface

In recent years, the technology of optical communication has gained importance due to high bandwidth and data rate requirements. This book focuses on free-space optical (FSO) communication that is capable of providing cable-free communication at very high data rates (up to Gbps). Unlike radio frequency communication that has restricted bandwidth due to its limited spectrum availability and interference, FSO communication has license-free spectrum as of now. This technology finds its application in terrestrial links, deep space/inter-satellite links, unmanned aerial vehicles (UAVs), high-altitude platforms (HAPs), and uplink and downlink between space platform, aircrafts, and other ground- based fixed/mobile terminals. It provides good privacy with flexible interconnection through a distributed or centralized communication system. It is a growing area of research these days due to its low power and mass requirement, bandwidth scalability, unregulated spectrum, rapid speed of deployment/redeployment, and cost-effectiveness. However, despite many advantages, the performance of FSO communication system is influenced by unpredictable atmospheric conditions, and this undoubtedly poses a great challenge to FSO system designers. The primary factors that deteriorate the FSO link performance are absorption, scattering, and turbulence. Out of these, the atmospheric turbulence is a major challenge that may lead to serious degradation in the link performance and make the communication link infeasible. This book gives the basic understanding of FSO communication system and lays emphasis on improving the performance of FSO link in turbulent atmosphere.

The purpose of this book is to cover the basic concepts of FSO communication system and provide the readers with sufficient in-depth knowledge to design a wireless optical link. The intended readers for this book include engineers, designers, or researches who are interested in understanding the phenomena of laser beam propagation through the atmosphere. This book primarily focuses on outdoor wireless communication, though a little briefing on indoor wireless communication is given in the introductory chapter. Although this book is based on the doctoral work of the first author, it has been completely rewritten and expanded to cover basic concepts of FSO communication system from readers' point of view.

This book has been organized into seven chapters. Chapter 1 provides an overview of FSO technology with historical background and its various applications. Chapter 2 gives a comprehensive coverage of FSO channel models and various atmospheric losses encountered during beam propagation through the atmosphere including free-space loss, pointing loss, absorption, and scattering loss. This is followed by the description of atmospheric turbulence and its effects on the laser communication, i.e., beam wander, beam spreading, beam scintillation, spatial coherence degradation, and image dancing. Various models for the atmospheric turbulent channel are presented. Chapter 3 discusses various components of FSO communication system. It provides description of optical transmitter, amplifiers, and receiver. The design of optical receiver that takes into account different types of detectors, noise sources, and receiver performance in terms of signal-to-noise ratio is presented. Finally, various issues involved in the link design like choice of operating wavelength, aperture diameter, and receiver bandwidth are discussed. Chapter 4 deals with the most challenging aspect of FSO communication system, i.e., acquisition, tracking, and pointing. The initial linkup or acquisition time puts a limit on the overall performance of the system, and hence, it is an essential system design constraint. Various subsystems involved in the accurate pointing of narrow laser beam toward the target are presented in this chapter. Chapter 5 presents bit error rate (BER) performance of FSO link for coherent and noncoherent modulation schemes. Chapter 6 discusses various techniques for improving link performance, i.e., aperture averaging, spatial diversity, coding, adaptive optics, relay-assisted FSO, etc. Finally, the last chapter describes in detail how the optical system designers can calculate link budgets.

Gurgaon, India — Hemani Kaushal
New Delhi, India — V.K. Jain
New Delhi, India — Subrat Kar

Contents

List of Figures

List of Tables

List of Symbols

θ_s	Planar emission angle
α_a	Aerosol absorption coefficient
α_m	Molecular absorption coefficient
α_r	Angular pointing error
α_T	Transmitter truncation ratio
β	Modulation index
β_a	Aerosol scattering coefficient
$\beta_{fog}(\lambda)$	Specific attenuation of fog
β_m	Molecular scattering coefficient
$\Delta\lambda_{filter}$	Bandwidth of optical band pass filter
ϵ_T	Root sum square of two-axis pointing bias error
η	Quantum efficiency of the detector
η_λ	Narrow-band filter transmission factor
η_R	Receiver optics efficiency
η_{TP}	Transmitter pointing loss factor
η_T	Transmitter optics efficiency
γ	Atmospheric attenuation coefficient
Γ_2	Mutual coherence function of second order
Γ_{code}	Coding gain
γ_i	Instantaneous SNR
γ_R	Receiver obscuration ratio
γ_s	Scattering angle
γ_T	Transmitter obscuration ratio
κ	Scalar spatial frequency
Λ	Receiver beam parameter (amplitude change due to diffraction)
λ	Operating wavelength
λ_0	Transmitter beam parameter (amplitude change due to diffraction)
λ_B	Rate of arrival of background photons
λ_s	Rate of arrival of signal photons
$\langle r_c^2 \rangle$	Beam wander displacement variance
$\mathbb{L}$	Constraint length of code

$\mathbb{P}$	Peak-to-average power ratio of the signal
$\mathbf{R}$	Rainfall rate
$\mathbf{V}$	Characteristic velocity
$\mathcal{F}$	Fresnel length
$\mathcal{M}$	Avalanche multiplication factor
$\mathtt{r}$	Radius of atmospheric particles
$\mathtt{h}$	Planck's constant
ν	Operating frequency
ν_k	Kinematic viscosity
Ω_b	Beam solid angle
Ω_{FOV}	Solid angle receiver field of view
ω_{IF}	Intermediate frequency
ω_L	Frequency of local oscillator
Ω_S	Stellar or point source field of view
Ω_s	Emission angle
ω_s	Frequency of incoming signal
ϕ	Phase of transmitted signal
Φ_n	Power spectral density of refractive index fluctuations
Ψ	Complex phase fluctuations
ρ	Correlation among beams
σ_b^2	Background noise current variance
σ_d^2	Detector dark current noise variance
σ_I^2	Scintillation index
σ_l^2	Variance of log-irradiance
σ_{pe}	Effective pointing error displacement
σ_R^2	Rytov variance
σ_s^2	Signal shot noise variance
σ_{Th}^2	Thermal noise variance
σ_{tilt}	RMS turbulence-induced wavefront tip/tilt
σ_T	Root sum square of two-axis jitter
σ_x^2	Variance of large-scale irradiance fluctuations
σ_y^2	Variance of small-scale irradiance fluctuations
$\sigma_{\ln x}^2$	Variance of large-scale log-irradiance
$\sigma_{\ln y}^2$	Variance of small-scale log-irradiance
τ	Optical depth
Θ	Receiver beam parameter (amplitude change due to refraction)
θ	Zenith angle
Θ_0	Transmitter beam parameter (amplitude change due to refraction)
θ_0	Isoplanatic angle
θ_{div}	Beam divergence
θ_{FOV}	Angular field of view of receiver
θ_H	Azimuth pointing error angle
θ_{jitter}	Beam jitter angle
θ_{unc}	Area of uncertainty in solid angle

θ_V	Elevation pointing error angle
$\triangle f_c$	Coherence bandwidth
$\triangle t_c$	Coherence time
ε	Overlap factor
ξ	Normalized distance variable
ξ_t	Safety margin against high-frequency fluctuations
A	Photodiode area
A_0	Amplitude of Gaussian beam
A_f	Aperture averaging factor
A_R	Effective area of the receiver
A_s	Surface area
B	Signal bandwidth
B_d	Doppler spread
B_o	Optical filter bandwidth
C	Channel capacity
c	Velocity of light
C_n^2	Refractive index structure constant
C_t^2	Temperature structure constant
C_v	Velocity structure constant
D	OFDM bias component
D_R	Receiver aperture diameter
D_t	Structure function for temperature
D_n	Structure function for refractive index
D_v	Structure function for wind velocity
E_{LO}	Local oscillator signal voltage
E_R	Received signal voltage
e_L	Electric field of local oscillator
e_s	Electric field of incoming signal
F	Excess noise factor
f	Signal frequency
F'	Phase front radius of curvature of the beam at the receiver plane
F_0	Phase front radius of curvature of the beam at the transmitter plane
F_n	Noise figure
G_R	Receiver gain
G_T	Transmitter gain
H	Altitude of the satellite
h	Plank's constant
h_0	Altitude of the transmitter
H_B	Background radiance of extended sources
I	Irradiance/intensity
I_0	Irradiance without turbulence
I_λ	Exo-atmospheric solar constant
I_{BG}	Background noise current
I_{db}	Bulk dark current
I_{ds}	Surface dark current

I_d Dark current
I_p Photodetector current
k Wave number
K_B Boltzmann's constant
K_b Average number of noise photons
k_b Number of information or data bits
k_{eff} Ionization ratio
K_s Average number of signal photons
L_0 Turbulent eddy outer scale size
l_0 Turbulent eddy inner scale size
l_f Dimension of turbulent flow
L_G Beam divergence loss
L_p Pointing loss
L_R Transmission loss of receiver optics
L_s Space loss factor
m Number of memory registers
N Number of receivers
n Index of refraction
n_0 Mean value of index of refraction
N_B Irradiance energy densities of point sources
n_c Length of code
N_r Number of total receiver scan area repeats
n_{sp} Spontaneous emission factor
N_t Number of total transmitter scan area repeats
P' Atmospheric pressure
P_{acq} Probability of acquisition
P_B Background noise power
P_{ce} Probability of chip error
$P_{detection}$ Probability of detection
P_{ew} Probability of word error
P_e Probability of error
P_L Power of local oscillator
P_R Received power
P_{sp} Amplifier spontaneous output noise power
P_s Power of incoming signal
P_T Transmitted power
q Electronic charge
R Link range
r Spatial separation of two points in space
r_0 Atmospheric coherence length
R_b Bit rate
R_{dwell} Receiver dwell time
R_L Load resistance
Re Reynolds number
S_n Noise power spectral density

T	Absolute temperature in Kelvin
T'	Atmospheric temperature
T_θ	Transmittance factor
T_a	Atmospheric transmittance
T_b	Bit duration
T_{dwell}	Transmitter dwell time
T_m	Multipath spread
T_{SS}	Single scan acquisition time
T_{ss}	Beam spread due to atmospheric turbulence
T_s	Slot width
U	Electric field
W	Effective beam radius at the receiver
W_0	Transmitter beam size
w_c	Number of 1s in each column in sparse matrix
W_e	Effective spot size in turbulence
W_{LT}	Long-term spot size
w_r	Number of 1s in each row in sparse matrix
p	Size distribution coefficient of scattering
V	Visibility range

List of Abbreviations

AF	Amplify-and-Forward
AM	Amplitude Modulation
AO	Adaptive Optics
APD	Avalanche Photodetector
ASE	Amplified Spontaneous Emission
ATP	Acquisition, Tracking, and Pointing
AWGN	Additive White Gaussian Noise
BER	Bit Error Rate
BPSK	Binary Phase Shift Keying
BSTS	Boost Surveillance and Tracking System
CALIPSO	Cloud-Aerosol Lidar and IR Pathfinder Satellite Observation
CCD	Charge-Coupled Devices
CDF	Cumulative Distribution Function
CF	Compress-and-Forward
DAPIM	Differential Amplitude Pulse Interval Modulation
DAPPM	Differential Amplitude Pulse Position Modulation
DEF	Detect-and-Forward
DHPIM	Dual Header Pulse Interval Modulation
DOLCE	Deep Space Optical Link Communications Experiment
DPIM	Differential Pulse Interval Modulation
DPPM	Differential Pulse Position Modulation
EGC	Equal-Gain Combining
ESA	European Space Agency
ETS	Engineering Test Satellite
FDM	Frequency Division Multiplexing
FIR	Far-Infrared
FM	Frequency Modulation
FOU	Field of Uncertainty
FOV	Field of View
FPA	Focal Pixel Array
FSO	Free-Space Optical

FSOI	FSO Interconnect
GOLD	Ground/Orbiter Lasercomm Demonstration
GOPEX	Galileo Optical Experiment
HAP	High-Altitude Platform
IF	Intermediate Frequency
IM/DD	Intensity Modulated/Direct Detection
IR	Infrared
ISRO	Indian Space Research Organisation
JPL	Jet Propulsion Laboratory
KIODO	KIrari's Optical Downlink to Oberpfaffenhofen
LCS	Laser Cross-Link Subsystem
LD	Laser Diode
LDPC	Low-Density Parity Check
LED	Light-Emitting Diode
LIR	Long-Infrared
LO	Local Oscillator
LOLA	Airborne Laser Optical Link
LOS	Line-of-Sight
LPF	Low-Pass Filter
MEMS	Microelectromechanical System
MIR	Mid-infrared
MISO	Multiple Input Single Output
MLCD	Mars Laser Communication Demonstration
MLSD	Maximum Likelihood Sequence Detection
MOLA	Mars Orbiter Laser Altimeter
MRC	Maximum-Ratio Combining
NASA	National Aeronautics and Space Administration
NBF	Narrow-Band Filter
NEA	Noise Equivalent Angle
NIR	Near-Infrared
NRZ	Non-return to Zero
NSDA	National Space Development Agency
OICETS	Optical Inter-orbit Communications Engineering Test Satellite
OOK	On-Off Keying
OTG	Optical Turbulence Generator
PAA	Point Ahead Angle
PAM	Pulse Amplitude Modulation
PAPM	Pulse Amplitude and Pulse Position Modulation
PAPR	Peak-to-Average Power Ratio
PCB	Printed Circuit Board
PDF	Probability Density Function
PER	Packet Error Rate
PPM	Pulse Position Modulation
QAM	Quadrature Amplitude Modulation
QAPD	Quadrant Avalanche Photodetector

QPIN	Quadrant P-Intrinsic
QPSK	Quadrature Phase Shift Keying
RF	Radio Frequency
ROSA	RF Optical System Study for Aurora
RSS	Root Sum Square
RZ	Return to Zero
SC	Selection Combining
SFTS	Space Flight Test System
SILEX	Space Intersatellite Link Experiment
SIR	Short-Infrared
SISO	Single Input Single Output
SNR	Signal-to-Noise Ratio
SOLACOS	Solid State Laser Communications in Space
SROIL	Short-Range Optical Intersatellite Link
TES	Tropospheric Emission Spectrometer
TPPM	Truncated PPM
UAV	Unmanned Aerial Vehicle
VLC	Visible Light Communication
VLSI	Very-Large-Scale Integration
WBAN	Wireless Body Area Network
WLAN	Wireless Local Area Network
WOC	Wireless Optical Communication
WPAN	Wireless Personal Area Network

Chapter 1
Overview of Wireless Optical Communication Systems

1.1 Introduction

WOC communication is considered as the next frontier for high-speed broadband connection due to its unique features: extremely high bandwidth, ease of deployment, tariff-free bandwidth allocation, low power ($\sim$1/2 of radio-frequency (RF) systems), less mass ($\sim$1/2 of RF systems), small size ($\sim$1/10 the diameter of RF antenna), and improved channel security. It has emerged a good commercial alternative to existing radio-frequency communication as it supports larger data rates and provides high gain due to its narrow beam divergence. It is capable of transmitting data up to 10 Gbps and voice and video communication through the atmosphere/free space. WOC have two broad categories, namely, indoor and outdoor wireless optical communications. Indoor WOC is classified into four generic system configurations, i.e., directed line-of-sight (LOS), non-directed LOS, diffused, and quasi diffused. Outdoor wireless optical communication is also termed as free-space optical (FSO) communication. The FSO communication systems are also classified into terrestrial and space systems. Figure 1.1 shows the classification of WOC systems.

Over the last few years, massive expansion in WOC technology has been observed due to huge advances in optoelectronic components and tremendous growth in the market offering wireless optical devices. It seems to be one of the promising technologies for addressing the problem of huge bandwidth requirements and "last mile bottleneck." There are many commercial applications of WOC technology which includes ground-to-LEO, LEO-to-GEO/LEO-to-ground, GEO-to-ground, LEO/GEO-to-aircraft, deep space probes, ground stations, unmanned aerial vehicles (UAVs), high-altitude platforms (HAPs), etc. [1–4]. It also finds applications in the area of remote sensing, radio astronomy, space radio communication, military, etc. When WOC technology is used over very short distances, it is termed as FSO interconnects (FSOI), and it finds applications in chip-to-chip or board-to-board interconnections. FSOI has gained popularity these days

H. Kaushal et al., *Free Space Optical Communication*, Optical Networks,
DOI 10.1007/978-81-322-3691-7_1

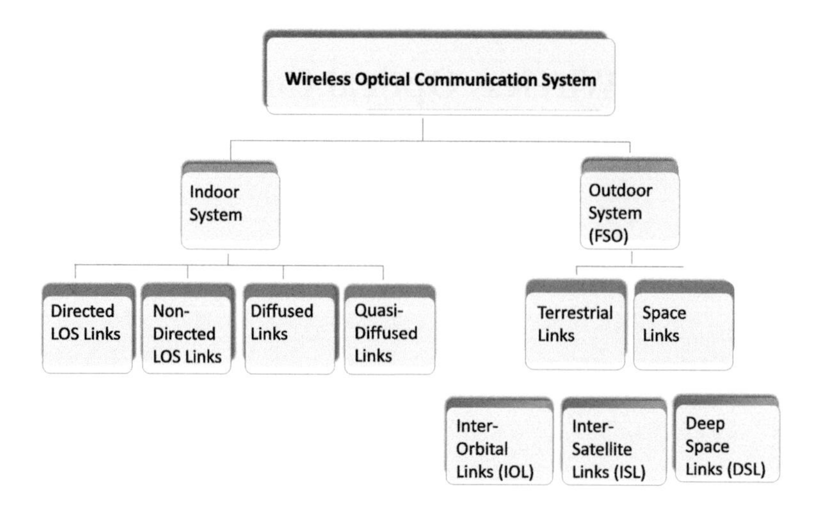

Fig. 1.1 Classification of wireless optical communication systems

as it potentially addresses complex communication requirement in optoelectronic devices. This technology offers the potential to build interconnection networks with higher speed, lower power dissipation, and more compact packages than possible with electronic very large-scale integration (VLSI) technology. However, the cost of optoelectronic devices, their integration, and overall packaging makes FSOI a costly affair. A throughput upto 1 Tbps per printed circuit board (PCB) has been experimentally demonstrated in [5] using 1000 channels per PCB with 1 mm optical beam array at 1 Gbps per channel.

Based on their transmission range, WOC can be classified into five broad categories (refer Fig. 1.2):

(i) Ultrashort-range WOC – used in chip-to-chip communication or all optical lab-on-a-chip system.
(ii) Short-range WOC – used in wireless body area networks (WBANs) or wireless personal area networks (WPANs).
(iii) Medium-range WOC – used in indoor IR or visible light communication (VLC) for wireless local area networks (WLANs) and inter-vehicular and vehicle-to-infrastructure communications.
(iv) Long-range WOC – used in terrestrial communication between two buildings or metro area extensions.
(v) Ultra-long-range WOC – used in ground-to-satellite/satellite-to-ground or inter-satellite link or deep space missions.

Commercially available FSO equipment provide much higher data rates ranging from 10 Mbps to 10 Gbps [10, 11]. Many optical companies like LightPointe in San Diego, fSONA in Canada, CableFree Wireless Excellence in UK, AirFiber in

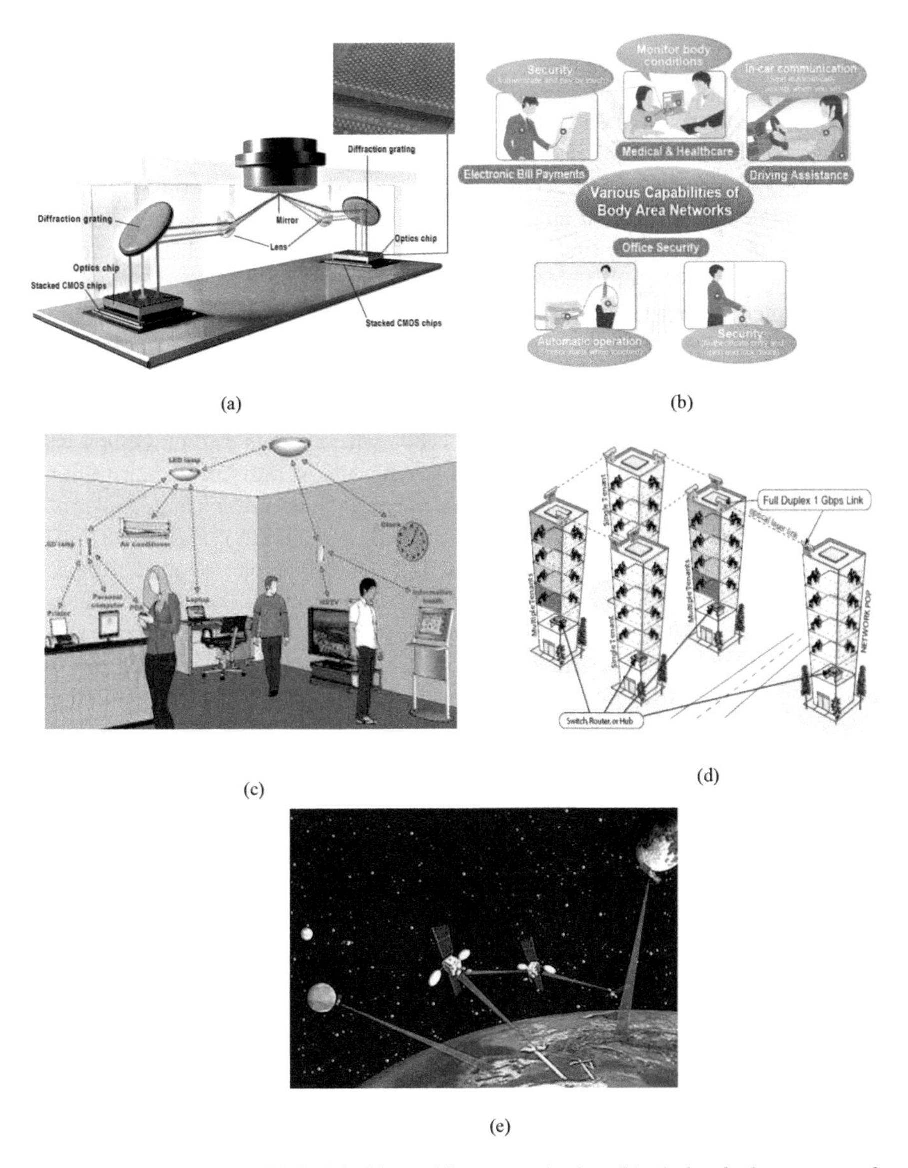

Fig. 1.2 Applications of WOCs: (**a**) chip-to-chip communication, (**b**) wireless body area network, (**c**) indoor IR or visible light communication, (**d**) inter-building communication, and (**e**) deep space missions [6–9]

California, etc. provide a wide range of wireless optical routers, optical wireless bridges, hybrid wireless bridges, switches, etc. that can support enterprise connectivity, last mile access, and HDTV broadcast link with almost 100 % reliability in adverse weather conditions.

1.1.1 History

The first experiment of transmitting signal over the atmosphere was conducted by Alexander Graham Bell in 1880. He used sunlight as a carrier to transmit voice signal over a distance of about a few feet. However, the experiment was not successful due to inconsistent nature of the carrier. Later, in the 1960s, Theodore H. Maiman discovered the first working laser at Hughes Research Laboratories, Malibu, California. From this point onward, the fortune of FSO has changed. Various experiments were conducted in military and space laboratories to demonstrate FSO link. In the 1970s, the Air Force sponsored a program known as *Space Flight Test System* (SFTS) to establish satellite to ground link at Air Force ground station, New Mexico. The program was later renamed as *Airborne Flight Test System*. This program achieved its first success in the 1980s where a data rate of 1 Gbps was demonstrated from aircraft to ground station. After this, a flurry of demonstrations were recorded during the 1980s and 1990s. They include *Laser Cross-Link Subsystem* (LCS), *Boost Surveillance and Tracking System* (BSTS), *Follow-On Early Warning System* (BSTS), and many more [12]. A full duplex ground to space laser link known as *Ground/Orbiter Laser Communication Demonstration* was first established in 1995–1996 by National Aeronautics and Space Administration (NASA) in conjunction with Jet Propulsion Laboratory (JPL). In addition to this, various demonstrations were carried out for deep space and inter-satellite missions such as *Mars Laser Communication Demonstration* (MLCD) and *Space Intersatellite Link Experiment* (SILEX) [12], respectively.

Very large-scale development is carried out by NASA in the USA, Indian Space Research Organisation (ISRO) in India, European Space Agency (ESA) in Europe, and National Space Development Agency (NSDA) in Japan. Demonstrations have established a full duplex FSO link with high data rates between various onboard space stations and ground stations, inter-satellite, etc. with improved reliability and 100 % availability. Besides FSO uplink/downlink, extensive research is carried out for FSO terrestrial links, i.e., link between two buildings to establish local area network segment that will provide last mile connectivity to the users (Fig. 1.3).

FSO communication is well suited for densely populated urban areas where digging of roads is cumbersome. Terrestrial FSO links can be used either for short range (few meters) or long range (tens of km). Short-range links provide high-speed connectivity to end users by interconnecting local area network segments that are housed in building separated within the campus or different building of the company. Long-range FSO communication links extend up to existing metropolitan area fiber rings or to connect new networks. These links do not reach the end user but they extend their services to core infrastructure. FSO communication system can also be deployed within a building, and it is termed as *indoor wireless optical communication (WOC) system*. This short-range indoor WOC system is a futuristic technology and is gaining attention these days with the advancement of technology involving portable devices, e.g., laptops, personal digital assistants, portable telephonic devices, etc.

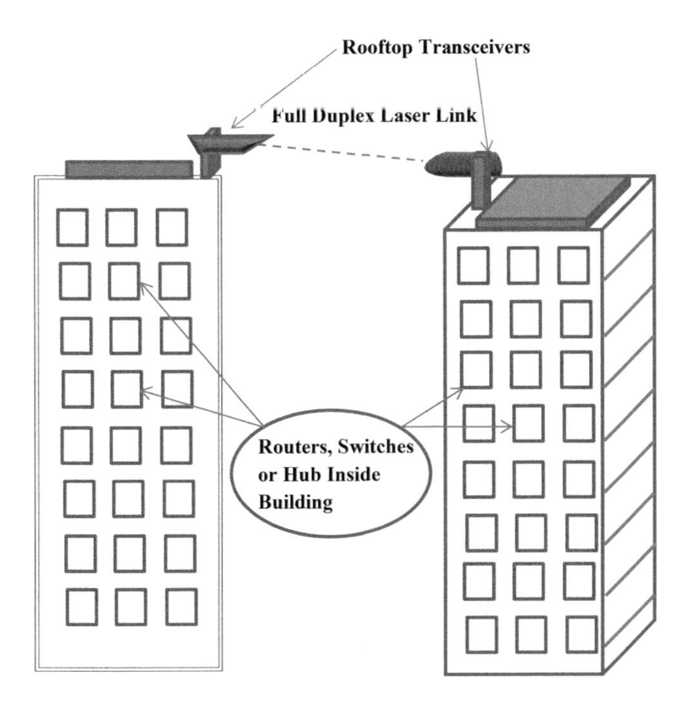

Fig. 1.3 FSO terrestrial link [13]

1.1.2 Indoor Wireless Optical Communication

Indoor wireless optical communication links provide a flexible interconnection within a building where setting up a physical wired connection is cumbersome. It consists of lasers or light-emitting diodes as transmitter and photodetectors as the receiver. These devices along with their drive circuits are much cheaper as compared to radio-frequency equipment or existing copper cables. Further, indoor WOC is inherently secure technology since the optical signals do not penetrate walls unlike electromagnetic waves which can cause interference and thus provides a high degree of security against eavesdropping. These optical waves are either in the visible light spectrum or in the IR spectrum which is able to provide very large (THz) bandwidth. Since these devices consume very little power, they are also suitable for mobile terminal systems. Besides many advantages, indoor optical wireless system is influenced by various impairments that impact the performance of the communication system. Some of the factors that lead to these impairments are (i) limiting speed of optoelectronic devices; (ii) large path loss; (iii) noisy indoor environment due to incandescent, fluorescent lighting or sunlight that contributes to noise in the detector; (iv) multipath dispersion; and (v) interference due to artificial noise sources. The range of the system is restricted as the average transmitted power is limited due to eye safety regulations [14].

The most commonly used optical sources in IR transmitters for indoor WOC are light-emitting diodes (LEDs) and laser diodes (LDs). LEDs are preferred over LDs as they are cheap and have broader modulation bandwidth and linear characteristics in the operating region. Since LEDs are nondirectional optical sources, their output power is not very high. Therefore, to compensate for lower power levels, an array of LEDs can be used. However, LEDs cannot work at high data rates beyond 100 Mbps. Laser diodes can be used at data rates of the order of few Gbps. Due to eye safety regulations, laser diode cannot be used directly for indoor WOC as they are highly directional and can cause optical damage.

1.1.2.1 Types of Link Configurations

The classification of an indoor optical link depends upon two major factors: (i) transmitter beam angle, i.e., degree of directionality, and (ii) the detector' s field-of-view (FOV), i.e., whether the view of the receiver is wide or narrow. Based on this, there are mainly four types of link configuration, i.e., directed line-of-sight (LOS), non-directed LOS, diffused, and multi-beam quasi diffused links.

(i) **Directed LOS Link**: In this type of link, the beam angle of transmitter as well as FOV of receiver are very narrow. The transmitter and receiver are directed toward each other. This configuration is good for point-to-point link establishment for indoor optical communication. The advantages and disadvantages of directed LOS link are as follows:

Advantages:

- Improved power efficiency as path loss is minimum
- Reduced multipath distortion
- Larger rejection of ambient background light
- Improved link budget

Disadvantages:

- Links are highly susceptible to blocking (or shadowing), and therefore, they cannot provide mobility in a typical indoor environment.
- Reduced flexibility as it does not support point-to-multipoint broadcast links.
- Tight alignment between transmitter and receiver is required making it less convenient for certain applications.

Figure 1.4 shows the pictorial representation of the link. This configuration has been used for many years at low data rate for electronic appliances using remote control applications like television or audio equipments. It provides point-to-point connectivity between portable electronic devices such as laptops, mobile equipments, PDAs, etc. Depending upon the degree of directionality, there is another version of LOS link called hybrid LOS. In this case, transmitter and receiver are facing each other, but the divergence

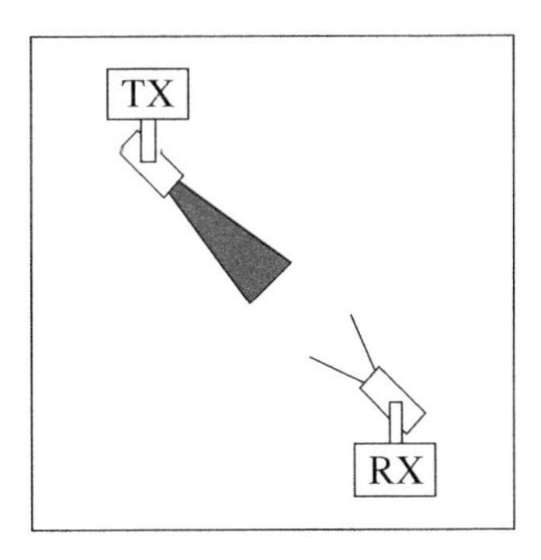

Fig. 1.4 Directed LOS link [14]

angle of transmitter is much larger than FOV of receiver. This configuration provides larger coverage area than directed LOS, but at the cost of reduced power efficiency, and it also suffers from blocking problems.

(ii) **Non-directed LOS Link:** In this type of link, the beam angle of transmitter and detector FOV is wide enough to ensure the coverage throughout the indoor environment. Such links do not require tight pointing and alignment as compared to directed LOS. However, in this case, the received irradiance is reduced for a given link distance and transmitted power. This link is suited for point-to-multipoint broadcast applications since it provides the desired high degree of mobility. In case of larger room dimensions, the entire room can be divided into multiple optical cells, and each cell is controlled by a separate transmitter with controlled beam divergence. The advantages and disadvantages of this link configuration are as follows:

Advantages:

- Allow high user mobility
- Increased robustness against shadowing
- Alleviate the need of pointing
- Well suited for point-to-multipoint broadcast applications

Disadvantages:

- The received signal suffers from multipath distortion as the beam gets reflected from walls or other objects in the room due to wider beam divergence
- Less power efficient

This link provides connectivity up to tens of meters and supports data rate up to 10 Mbps which can be shared among users operating within the cell. Figure 1.5 depicts a non-directed LOS link.

(iii) **Diffused Link:** In this type of link, the transmitter is facing the ceiling/roof, and it emits a wide beam of IR energy toward the ceiling. The IR signal after it undergoes multiple reflections from walls or room objects is collected by a receiver placed on the ground with wider FOV. Its advantages and disadvantages are as follows:

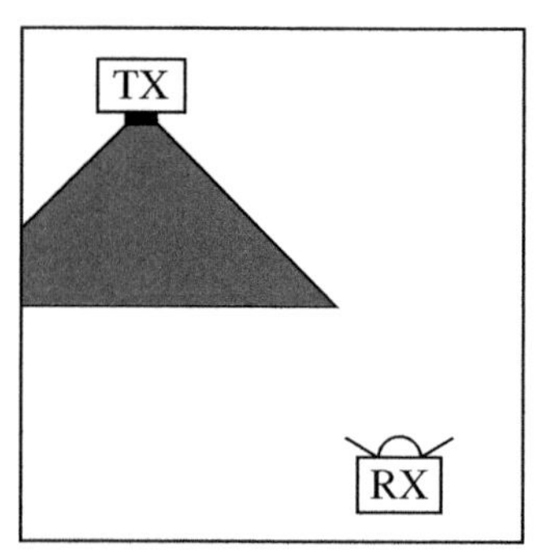

Fig. 1.5 Multi-beam non-directed LOS link [14]

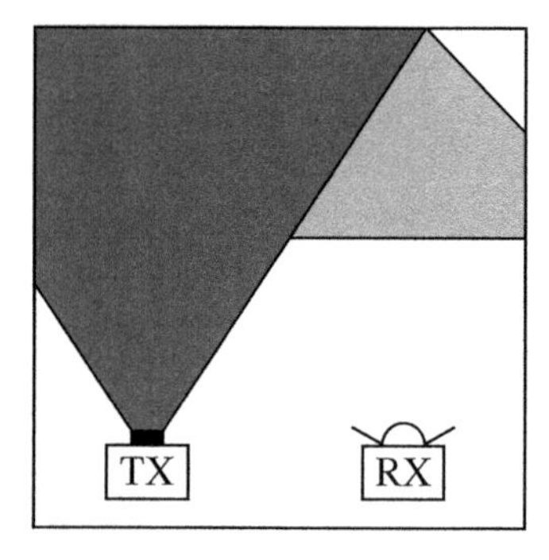

Fig. 1.6 Diffused link [14]

Advantages:

- There is no requirement of alignment between transmitter and receiver as the optical signal is uniformly spread within the room by making use of reflective properties of walls and ceilings.
- This link is most robust and flexible as it is less prone to blocking and shadowing.

Disadvantages:

- Severe multipath distortion
- High optical path loss typically 50–70 dB for a link range of 5 m [15].

Diffuse link is the link configuration of choice for IEEE 802.11 infrared physical layer standard. It can support data rates up to 50 Mbps and is typically as shown in Fig. 1.6.

(iv) **Multi-Beam Quasi Diffused Link:** In this type of link, a single wide beam diffuse transmitter is replaced by multi-beam transmitter, also known as quasi diffused transmitter. The multiple narrow beams are pointed outward in different directions. These optical signals are collected by angle diversity receiver [16, 17] placed on the ground. Angle diversity of the receiver can be achieved in two ways: First is to employ multiple non-imaging receiving elements oriented in different directions, and each element is having its own lensing arrangement/concentrators. The purpose of the lensing arrangement or the concentrators is to improve the collection efficiency by transforming the light rays incident over a large area into a set of rays that emerge from a smaller area. This allows the usage of smaller photodetectors with lesser cost and improved sensitivity. However, this approach is not a good choice as it will make the receiver configuration very bulky. Second is by using the improved version of angle diversity, also called "fly-eye receiver" [18]. It consists of imaging optical concentrator with a segmented photodetector array placed at its focal plane. In both the cases, the photocurrent generated by the individual receiver is amplified and processed using various combining techniques. Various advantages and disadvantages of this link are [17]:

Advantages:

- Provides high optical gain over wide FOV
- Reduced effect of ambient light source
- Reduced multipath distortion and co-channel interference
- Immunity against blockage near receiver
- Reduced path loss

Disadvantages:

- Complex to implement
- Costly

In Fig. 1.7 quasi diffused links with multiple and single lens arrangement are depicted. Experimental studies have shown that this link can support data rate up to 100 Mbps.

Various organizations have implemented indoor WOC systems. The chronology of indoor optical wireless research is given in Table 1.1.

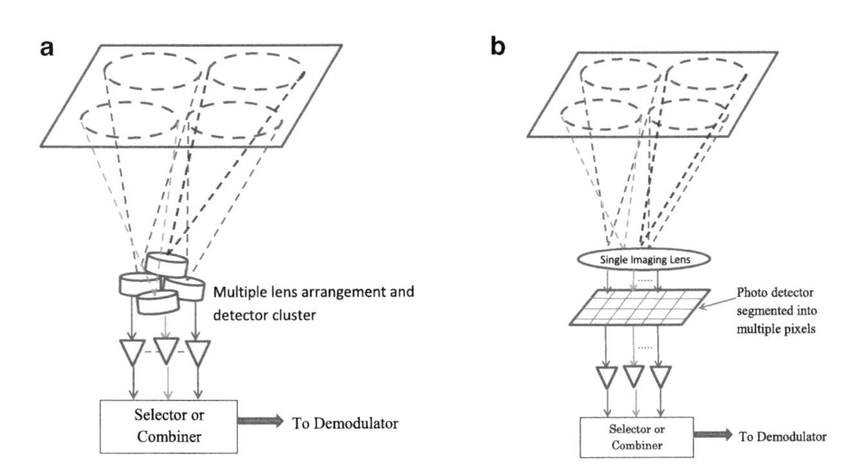

Fig. 1.7 Multi-beam quasi diffused links. (**a**) Receiver with multiple lens arrangement. (**b**) Receiver with single lens arrangement

Table 1.1 Chronology of indoor optical wireless research [19]

Date	Organization	Configuration	Bit rate	Characteristics
1979/1981	IBM	Diffuse	64–125 kbps	100 mW, 950 nm, BPSK
1983	Fujitsu	LOS	19.2 kbps	15 mW, 880 nm, FSK
1985	Hitachi	Hybrid	0.25–1 Mbps	300 mW, FSK
1985	Fujitsu	Hybrid	48 kbps	880 nm, BPSK
1987	Bell Labs	Directed LOS	45 Mbps	1 mW, 800 nm, OOK
1988	Matsushita	Hybrid	19.2 kbps	880 nm, FSK
1992	MPR Teltech Ltd	Non-directed	230 kbps	DPSK, 800/950 nm
1994	Berkeley	Diffused	50 Mbps	475 mW, 806 nm, OOK

Table 1.2 Comparison of indoor WOC and Wi-Fi systems

S.No.	Property	Wi-Fi radio	IR/VLC	Implication for IR/VLC
1	Spectrum licensing	Yes	No	Approval not required
				World wide compatibility
2	Penetration through walls	Yes	No	Inherently secure
				Carrier reuse in adjacent rooms
3	Multipath fading	Yes	No	Simple link design
4	Multipath dispersion	Yes	Yes	Problematic at high data rates
5	Dominant noise	Other users	Background light	Short range

Most of the indoor wireless optical links operate in 780–950 nm range. For this reason, indoor wireless systems is also called IR systems. For directed and non-directed LOS, typically one LED is used that emits an average power of tens of mW. In case of diffused configuration, an array of LEDs oriented in different directions are used so as to provide flexibility in coverage area. These LEDs transmit power in the range of 100–500 mW. The comparison of indoor WOC and Wi-Fi systems is given in Table 1.2.

1.1.3 Outdoor/Free-Space Optical Communication

Free-space optical communication requires line-of-sight connection between transmitter and receiver for propagation of information from one point to another. Here, the information signal from the source is modulated on the optical carrier, and this modulated signal is then allowed to propagate through the atmospheric channel or free space, rather than guided optical fibers, toward the receiver. Ground-to-satellite (optical uplink) and satellite-to-ground (optical downlink) involve propagation of optical beam through the atmosphere as well as in free space. Therefore, these links are a combination of terrestrial and space links. Figure 1.8 illustrates the application areas of FSO links.

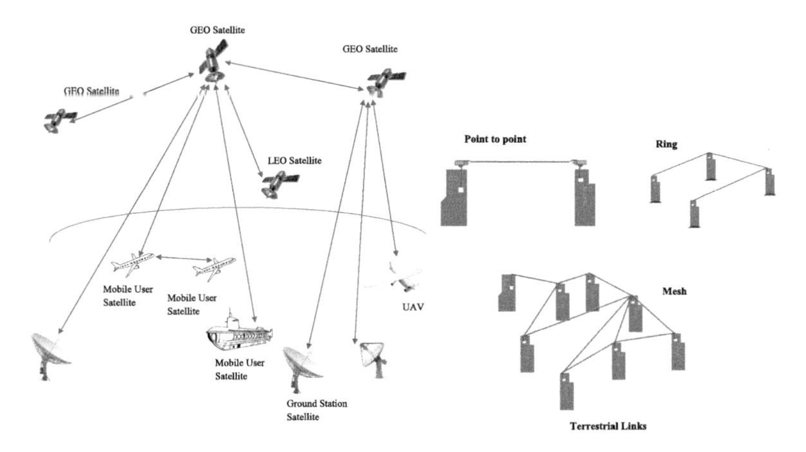

Fig. 1.8 Applications of FSO communication links

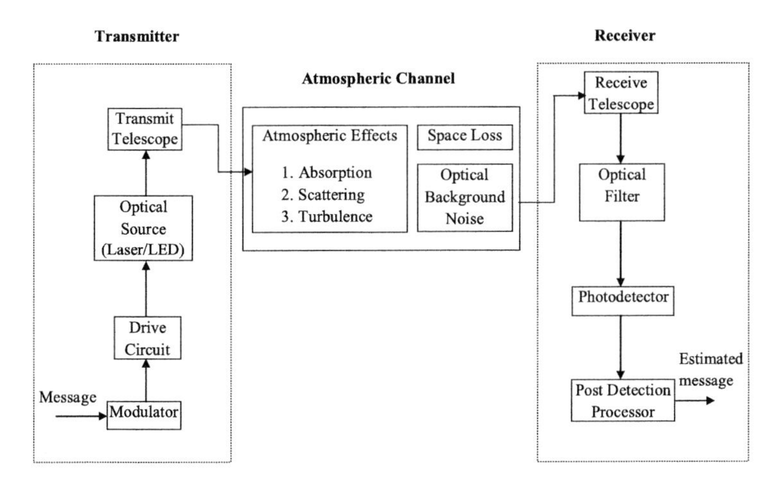

Fig. 1.9 Block diagram of FSO communication link

The basic block diagram of an FSO link is shown in Fig. 1.9. Like any other communication technologies, FSO communication link comprises of three basic subsystems, viz., transmitter, channel, and receiver [20].

(i) **Transmitter**: Its primary function is to modulate the message signal onto the optical carrier which is then propagated through the atmosphere to the receiver. The essential components of the transmitter are (a) the modulator, (b) the driver circuit for the optical source to stabilize the optical radiations against temperature fluctuations, and (c) the collimator or the telescope that collects, collimates, and directs the optical radiations toward the receiver. The most widely used modulation is the intensity modulation (IM) in which the source data is modulated on the irradiance/intensity of the optical carrier. This can be achieved by varying the driving current of the optical source directly with the message signal to be transmitted or by using an external modulator.

(ii) **Channel**: Since the FSO communication channel has the atmosphere as its propagating medium, it is influenced by unpredictable environmental factors like cloud, snow, fog, rain, etc. These factors do not have fixed characteristics and cause attenuation and deterioration of the received signal. The channel is one of the limiting factors in the performance of FSO system.

(iii) **Receiver**: Its primary function is to recover the transmitted data from the incident optical radiation. It consists of a receiver telescope, optical filter, photodetector, and demodulator. The receiver telescope collects and focuses the incoming optical radiation onto the photodetector. The optical filter reduces the level of background radiation and directs the signal on the photodetector that converts the incident optical signal into an electrical signal.

1.2 Comparison of FSO and Radio-Frequency Communication Systems

FSO communication system offers several advantages over the RF system. The major difference between FSO and RF communications arises from the large difference in the wavelength. Under clear weather conditions (visibility >10 miles), the atmospheric transmission window is in the near IR region and lies between 700 and 1600 nm. The transmission window for RF lies between 30 mm and 3 m. Therefore, RF wavelength is a thousand of times larger than the optical wavelength. This high ratio of wavelengths leads to some interesting differences between the two systems as given below:

(i) Huge modulation bandwidth: It is a well-known fact that an increase in carrier frequency increases the information-carrying capacity of a communication system. In RF and microwave communication systems, the allowable bandwidth can be up to 20 % of the carrier frequency. In optical communication, even if the bandwidth is taken to be 1 % of carrier frequency ($\approx 10^{16}$ Hz), the allowable bandwidth will be 100 THz. This makes the usable bandwidth at an optical frequency in the order of THz which is almost 10^5 times that of a typical RF carrier.

(ii) Narrow beam divergence: The beam divergence is $\sim\lambda/D_R$, where λ is the carrier wavelength and D_R the aperture diameter. Thus, the beam spread offered by the optical carrier is narrower than that of RF carrier. For example, the laser beam divergence at $\lambda = 1550$ nm and aperture diameter $D_R = 10$ cm come out to be 0.34 μrad. On the other hand, radio-frequency signal say at X band will produce a very large beam divergence, e.g., at 10 GHz i.e., $\lambda = 3$ cm, and aperture diameter $D_R = 1$ m yields beam divergence to be 67.2 mrad. Much smaller beam divergence at optical frequency leads to increase in the intensity of signal at the receiver for a given transmitted power. Figure 1.10 shows the

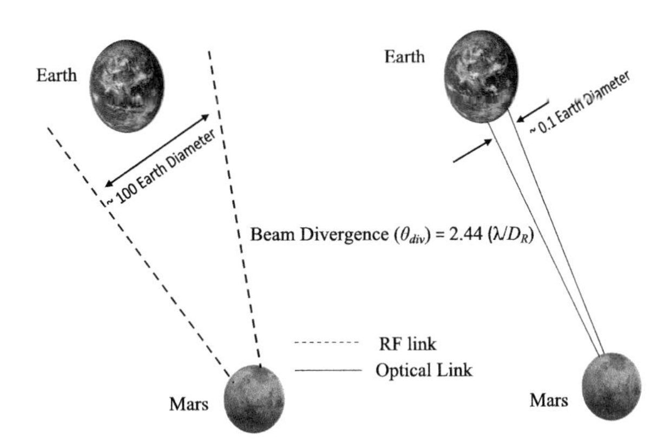

Fig. 1.10 Comparison of optical and RF beam divergence from Mars toward Earth [21]

comparison of beam divergence for optical and RF signals when sent back from Mars toward Earth [21].

(iii) Less power and mass requirement: For a given transmitter power level, the optical intensity is more at the receiver due to its narrow beam divergence. Thus, a smaller wavelength of optical carrier permits the FSO designer to come up with a system that has smaller antenna than RF system to achieve the same gain (as antenna gain is inversely proportional to the square of operating wavelength). The typical size for the optical system is 0.3 vs. 1.5 m for the spacecraft antenna [22].

(iv) High directivity: Since the optical wavelength is very small, a very high directivity is obtained with small-sized antenna. Antenna directivity is closely related to its gain. The advantage of optical carrier over RF carrier can be seen from the ratio of antenna directivity as given below

$$\frac{\text{Gain}_{\text{optical}}}{\text{Gain}_{\text{RF}}} = \frac{4\pi/\theta^2_{div(\text{optical})}}{4\pi/\theta^2_{div(\text{RF})}}, \tag{1.1}$$

where $\theta_{div(\text{optical})}$ and $\theta_{div(\text{RF})}$ are the optical and RF beam divergence, respectively, and are proportional to λ/D_R. For system using optical carrier with aperture diameter $D_R = 10$ cm and $\lambda = 1550$ nm gives $\theta_{div(\text{optical})} \approx 40\,\mu$rad. At beam divergence of 40 μrad, the antenna gain, $\text{Gain}_{\text{optical}}$, is approximately 100 dB. In order to achieve the same gain in RF system using X band at $\lambda = 3$ cm, the size of aperture diameter D_R becomes extremely large and unpractical to implement.

(v) Unlicensed spectrum: In RF system, interference from adjacent carrier is the major problem due to spectrum congestion. This requires the need of spectrum licensing by regulatory authorities. But optical system is free from spectrum licensing till now. This reduces the initial setup cost and development time.

(vi) Security: It is difficult to detect the transmuted optical beam as compared to RF signal because of its narrow beam divergence. In order to detect the transmitted optical signal, one has to be physically very close (≤ 0.1 miles) to the beam spot diameter. Studies have shown that the optical signal would drop to 140 dB from its peak transmission power at a distance of 10 miles. However, RF signal has much wider region of listening. In this case, signal can easily be picked up roughly at a distance of 40 miles and is down about 40 dB at approximately 100 miles.

In addition to the above advantages, the secondary advantages of FSO communication system are: (a) it is beneficial in the cases where fiber optic cables cannot be used, (b) easily expandable and reduces the size of network segments, and (c) light weight and compact.

Besides these advantages, an FSO communication system has got some disadvantages as well. It requires tight alignment and pointing due to its narrow beam divergence. Since light cannot penetrate walls, hills, buildings, etc., a clear line-of-sight is required between the transmitter and receiver. Also, unlike RF systems, an FSO system is highly prone to atmospheric conditions that can degrade the performance of system drastically. Another limiting factor is the position of the sun relative to the laser transmitter and receiver. In a particular alignment, solar background radiations will increase leading to poor system performance.

1.3 Choice of Wavelength in FSO Communication System

Wavelength selection in FSO communication system is a very important design parameter as it affects link performance and detector sensitivity of the system. Since antenna gain is inversely proportional to operating wavelength, it is more beneficial to operate at lower wavelengths. However, higher wavelengths provide better link quality and lower pointing-induced signal fade [23]. A careful optimization of operating wavelength in the design of FSO link will help in achieving better performance. The choice of wavelength strongly depends on atmospheric effects, attenuation, and background noise power. Further, the availability of transmitter and receiver components, eye safety regulations, and cost deeply impact the selection of wavelength in FSO design process.

The International Commission on Illumination [24] has classified optical radiations into three categories: IR-A (700–1400 nm), IR-B (1400–3000 nm), and IR-C (3000 nm–1 mm). It can be subclassified into:

(i) near-infrared (NIR) ranging from 750 to 1450 nm which is a low attenuation window and mainly used for fiber optics,
(ii) short-infrared (SIR) ranging from 1400 to 3000 nm out of which 1530–1560 nm is a dominant spectral range for long-distance communication,
(iii) mid-infrared (MIR) ranging from 3000 to 8000 nm which is used in military applications for guiding missiles,
(iv) long-infrared (LIR) ranging from 8000 nm to 15 μm which is used in thermal imaging, and
(v) far-infrared (FIR) which is ranging from 15 μm to 1 mm.

Almost all commercially available FSO systems are using NIR and SIR wavelength range since these wavelengths are also used in fiber-optic communication, and their components are readily available in the market.

The wavelength selection for FSO communication has to be eye and skin safe as certain wavelengths between 400 and 1500 nm can cause potential eye hazards or damage to the retina [25]. Table 1.3 summarizes various wavelengths used in practical FSO communication for space applications.

1.4 Range Equation for FSO Link

The link performance analysis begins with the basic component values and system parameters which are assumed to be known and are fixed in advance. For example, the laser with the fixed output power as a function of operating wavelength, transmitter, receiver telescope sizes, etc. directly affect the link performance. Therefore, the link performance analysis can be performed if all the components and operational parameters are precisely specified. The three basic steps for evaluating the performance of an optical link are as follows:

(i) Determine the number of detected signal photons at the detector taking into account the various losses in the transmitter, channel, and receiver.
(ii) Determine the number of detected background noise photons generated at the detector.
(iii) Compare the number of detected signal photons with the number of detected noise photons.

At the transmitter side, the optical source emits optical power with a varying degree of focusing often described by its emission angle. The total power (in watts) emitted from a uniform source characterized by brightness function B (watts/steradian·area), surface area A_s, and emission angle Ω_s is given by [43]

$$P_T = BA_s\Omega_s. \tag{1.2}$$

For symmetrical radiating sources, the solid emission angle Ω_s can be related to the planar emission angle θ_s (refer Fig. 1.11a) by

$$\Omega_s = 2\pi\left[1 - \cos\left(\theta_s/2\right)\right]. \tag{1.3}$$

For any Lambertian source that emits power uniformly in forward direction has $|\theta| \leq \pi/2$ implying $\theta_s = \pi$. It gives $\Omega_s = 2\pi$ and therefore transmitted power, $P_T = 2\pi BA_s$. The light can also be collected and refocused by means of beam-forming optics as shown in Fig. 1.11b. The light from the source beam is focused to a spot with the help of converging lens, and the diverging lens expands the beam to planar beam diameter D_R given by

Table 1.3 Wavelengths used in practical FSO communication system

Mission	Laser	Wavelength	Other parameters	Application
Semi-conductor Inter-satellite Link Experi-ment(SILEX) [26]	AlGaAs laser diode	830 nm	60 mW, 25 cm telescope size, 50 Mbps, 6 μrad divergence, direct detection	Inter-satellite comm.
Ground/Orbiter Lasercomm Demonstration (GOLD) [27]	Argon-ion laser/GaAs laser	Uplink: 514.5 nm Downlink: 830 nm	13 W, 0.6 m and 1.2 m tx. and rx.telescopes size, respectively; 1.024 Mbps, 20 μrad divergence	Ground-to-satellite link
RF Optical System Study for Aurora (ROSA) [28]	Diode pumped Nd:YVO4 laser	1064 nm	6 W, 0.135 m and 10 m tx. and rx. telescopes size, respectively; 320 kbps;	Deep space missions
Deep Space Optical Link Communications Experiment (DOLCE) [29]	Master oscillator power amplifier (MOPA)	1058 nm	1 W, 10–20 Mbps	Inter-satellite/Deep space missions
Mars Orbiter Laser Altimeter (MOLA) [30]	Diode pumped *Q* switched Cr:Nd:YAG	1064 nm	32.4 W, 420 μrad divergence, 10 Hz pulse rate, 618 bps; 850 μrad receiver field-of-view (FOV)	Altimetry
General Atomics Aeronautical Systems (GA-ASI) & TESAT [31]	Nd:YAG	1064 nm	2.6 Gbps	Remotely Piloted Aircraft (RPA) to LEO
Altair UAV-to-ground Lasercomm Demonstration [32]	Laser diode	1550 nm	200 mW, 2.5 Gbps, 19.5 μrad jitter error 10 cm and 1 m uplink and downlink telescopes size, respectively	UAV-to-ground link
Mars Polar Lander [33]	AlGaAs laser diode	880 nm	400 nJ energy in 100 nsec pulses, 2.5 kHz rate, 128 kbps	Spectroscopy
Cloud-Aerosol Lidar and IR Pathfinder Satellite Observation (CALIPSO) [34]	Nd:YAG	532 nm/1064 nm	115 mJ energy, 20 Hz rate, 24 ns pulse	Altimetry

(continued)

Table 1.3 (continued)

Mission	Laser	Wavelength	Other parameters	Application
KIrari's Optical Downlink to Oberpfaffenhofen (KIODO) [35]	AlGaAs laser diode	847/810 nm	50 Mbps, 40 cm and 4 m tx. and rx. telescopes size, respectively; 5 μrad divergence	Satellite-to-ground downlink
Airborne Laser Optical Link (LOLA) [36]	Lumics fiber laser diode	800 nm	300 mW, 50 Mbps	Aircraft and GEO satellite link
Tropospheric Emission Spectrometer (TES) [37]	Nd:YAG	1064 nm	360 W, 5 cm telescope size, 6.2 Mbps	Interferometry
Galileo Optical Experiment (GOPEX) [38]	Nd:YAG	532 nm	250 mJ, 12 ns pulse width, 110 μrad divergence, 0.6 m primary and 0.2 m secondary transmitter telescope size, 12.19 × 12.19 mm CCD array receiver	Deep Space Missions
Engineering Test Satellite VI (ETS-VI) [39]	AlGaAs laser diode (downlink) Argon laser (uplink)	Uplink: 510 nm Downlink: 830 nm	13.8 mW, 1.024 Mbps bidirectional link, direct detection, 7.5 cm spacecraft telescope size, 1.5 m Earth station telescope	Bi-directional ground-to-satellite link
Optical Inter-orbit Communications Engineering Test Satellite (OICETS) [40]	Laser Diode	819 nm	200 mW, 2.048 Mbps, direct detection, 25 cm telescope size	Bi-directional Inter-orbit link
Solid State Laser Communications in Space (SOLACOS) [41]	Diode pumped Nd:YAG	1064 nm	1 W, 650 Mbps return channel and 10 Mbps forward channel, 15 cm telescope size, coherent reception	GEO-GEO link
Short Range Optical Intersatellite Link (SROIL) [42]	Diode pumped Nd:YAG	1064 nm	40 W, 1.2 Gbps, 4 cm telescope size, BPSK homodyne detection	Inter-satellite link

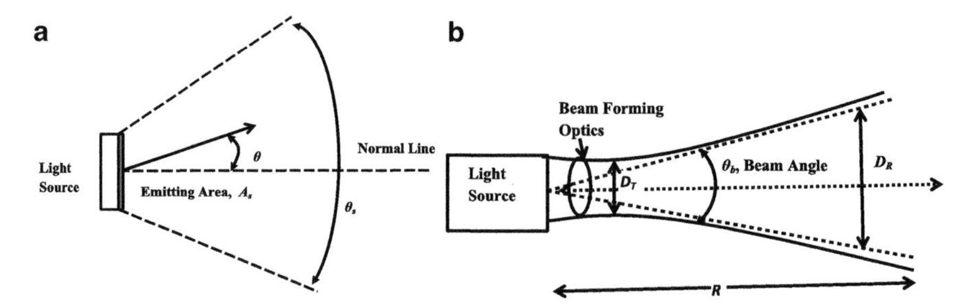

Fig. 1.11 Demonstration of optical emission from light source. (**a**) Light emission from Lambertian source [43]. (**b**) Light emission using beam forming optics [43]

$$D_R = D_T \left[1 + \left(\frac{\lambda R}{D_T^2} \right)^2 \right]^{1/2}, \tag{1.4}$$

where λ is the operating wavelength, D_T the transmitter lens diameter, and R the distance from the lens or link range.

$$\text{For} \begin{cases} \text{Near field, } \frac{\lambda R}{D_T^2} < 1,\ D_R \simeq D_T \\ \text{Far field, } \frac{\lambda R}{D_T^2} > 1,\ D_R \simeq \frac{\lambda R}{D_T} \end{cases} \tag{1.5}$$

First condition implies that emerging light is collimated with diameter equal to transmitter lens diameter. The second condition means that emerging light diverges with distance from the source. The planar beam angle, θ_b of the diverging light source, also called diffraction limited transmitter beam angle is approximately given by [43]

$$\theta_b \cong \frac{D_R}{R}. \tag{1.6}$$

Substituting the value of D_R for far field case, we get

$$\theta_b = \frac{\lambda}{D_T}. \tag{1.7}$$

The two-dimensional solid angle can be approximately related to planar beam angle by

$$\Omega_b = 2\pi \left[1 - \cos\left(\theta_b/2\right)\right] \cong \left(\frac{\pi}{4}\right) \theta_b^2. \tag{1.8}$$

Transmitter gain G_T from Eqs. (1.7) and (1.8) is given by

$$G_T = \frac{4\pi}{\Omega_b} \approx \left(\frac{4D_T}{\lambda}\right)^2. \tag{1.9}$$

After propagating through link distance R, the field intensity of the beam will be

$$I = \frac{G_T P_T}{4\pi R^2}. \tag{1.10}$$

A normal receiving area A within the beam collects the field power

$$P_R = \left(\frac{G_T P_T}{4\pi R^2}\right) A. \tag{1.11}$$

Let us define the receiver gain, G_R, in terms of A

$$G_R = \left(\frac{4\pi}{\lambda^2}\right) A => A = \frac{\lambda^2 G_R}{4\pi}. \tag{1.12}$$

Therefore, from Eqs. (1.11) and (1.12),

$$P_R = P_T G_T \left(\frac{\lambda}{4\pi R}\right)^2 G_R. \tag{1.13}$$

When the other loss factors are incorporated, the above equation becomes

$$P_R = P_T \left(G_T \eta_T \eta_{TP}\right) \left(\frac{\lambda}{4\pi R}\right)^2 \left(G_R \eta_R \eta_\lambda\right), \tag{1.14}$$

where

P_R : signal power at the input of photodetector
P_T : transmitter power
η_T and η_R : efficiencies of transmitter and receiver optics, respectively
G_T : gain of the transmitting antenna $\left[G_T \approx (4D_T/\lambda)^2\right]$
G_R : gain of the receiving antenna $\left[G_R \approx (4D_R/\lambda)^2\right]$
η_{TP} : transmitter pointing loss factor
$\left(\frac{\lambda}{4\pi R}\right)$: space loss factor, where R is the link distance
η_λ : narrowband filter transmission factor

From the above equation, it is seen that the receive signal power can be increased by one or more of the following options:

(i) Increasing transmit power: The most simplest way to improve the receive signal power is to increase the transmit power since the receive power scales linearly with the transmit power. However, increasing the transmit power implies the increase in the overall system power consumption, and this can lead to issues like safety, thermal management, etc.

(ii) Increasing transmit aperture: The transmit aperture size and beamwidth are inversely proportional to each other. Therefore, increasing the transmit aperture size will effectively reduce the transmitter beamwidth and hence deliver the signal with more intensity. However, it will lead to tight acquisition, pointing, and tracking requirement. Further, transmit aperture cannot be increased indefinitely as it will increase the overall mass of the terminal and that will increase the cost of the system.

(iii) Increasing receiver aperture: The receive signal power scales directly with the receive aperture area. However, the amount of background noise collected by the receiver will also increase with the increase in the receiver aperture area. It implies that the effective performance improvement does not always scale linearly with the receiver aperture area.

(iv) Reducing pointing loss: Reducing the transmitter and receiver pointing loss will improve the overall signal power level and will also reduce the pointing-induced signal power fluctuations.

(v) Improving overall efficiency: The overall efficiency can be improved by improving η_T, η_R, and η_λ through appropriate optics and filter design.

1.5 Technologies Used in FSO

Technologies used in FSO system are almost similar to conventional RF system. Most of the techniques are adapted directly from RF systems. In the following, various detection and modulation schemes used in FSO system are discussed. The modulation of optical carrier differs from the modulation of RF carrier because of the characteristics and limitations of the devices used for carrying the modulation process. The optical modulation can be carried out in two ways: internal or external as shown in Fig. 1.12. An internal modulator is one where the characteristics of the source are directly varied in accordance with the information signal to produce the modulated optical signal. Intensity modulation can be performed by varying the bias current. Frequency or phase modulation can be obtained by changing the cavity length of the laser. Pulse modulation can be achieved by varying the driving current above and below the threshold. These modulations are limited to the linear range of power characteristics of the source. In case of external modulators, an external device is used which varies the characteristics of the carrier in accordance with the modulating signal. These systems are capable of utilizing full power of the source. However, external modulators limit the modulation range and require relatively high drive current.

At optical frequencies, these modulators operate directly on carrier intensity (amplitude square of electric field) rather than the amplitude of the carrier. Other

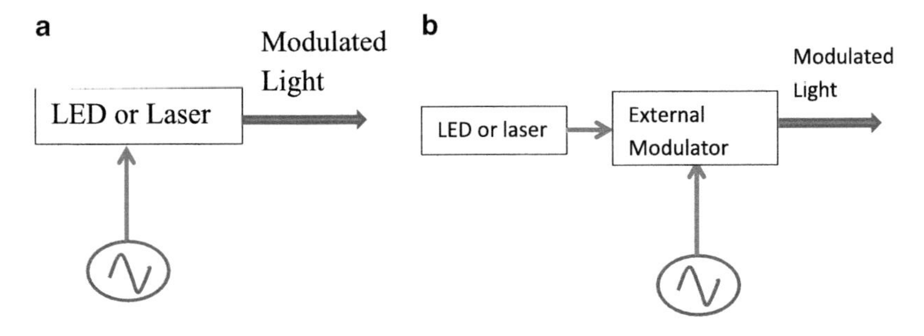

Fig. 1.12 Optical modulators. (**a**) Internal modulator. (**b**) External modulator

ways to modulate optical carrier are by using its phase or polarization. However, due to simplicity reasons, the most popular modulation scheme used in FSO system is intensity modulation: baseband or subcarrier. The modulation schemes can be classified into two categories: baseband intensity modulation and subcarrier intensity modulation. The most common method used for the detection of the optical signal is direct detection. When the intensity-modulated signal is detected by a direct detection receiver, the scheme is known as intensity-modulated/direct detection (IM/DD) and is commonly used in FSO systems. The other approach of detecting the modulated optical signal is coherent detection. It makes use of local oscillator to down convert the optical carrier to baseband (homodyne detection) or to RF intermediate frequency (heterodyne detection). This RF signal is subsequently demodulated to baseband via conventional RF demodulation process.

1.5.1 Direct Detection System

In direct detection technique, the received optical signal is passed through optical band-pass filter to restrict the background radiation. It is then allowed to fall on the photodetector which produces the output electrical signal proportional to the instantaneous intensity of the received optical signal. It may be regarded as linear intensity to current convertor or quadratic (square law) converter of optical electric field to detector current. The photodetector is followed by an electrical low-pass filter (LPF) with bandwidth sufficient enough to pass the information signal.

The signal-to-noise ratio (SNR) of direct detection receiver can be obtained by using noise models for a particular detector, i.e., PIN or avalanche photodetector (APD). With the received power given in Eq. (1.14) and detector noise sources, the SNR expressions are obtained. The SNR for PIN photodetector is given by

$$SNR = \frac{(R_0 P_R)^2}{2qB\,(R_0 P_R + R_0 P_B + I_d) + 4K_B TB/R_L}, \tag{1.15}$$

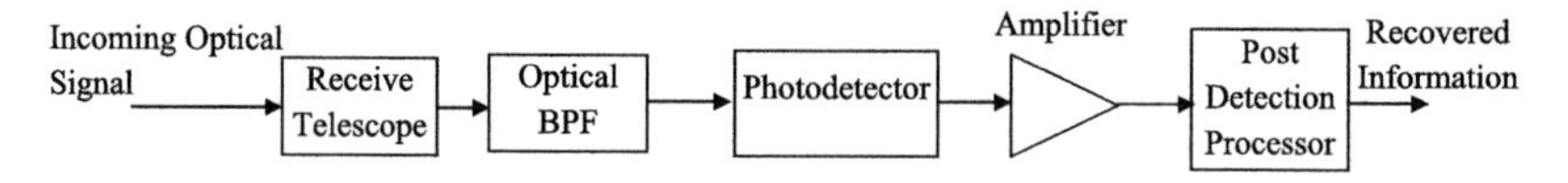

Fig. 1.13 Block diagram of direct detection receiver

where R_0 is the detector responsivity and is given by

$$R_0 = \frac{\eta q}{h\nu}. \tag{1.16}$$

In the above equation, η is the detector quantum efficiency, $q = 1.602 \times 10^{-19}$ J is the electronic charge, $h = 6.623 \times 10^{-34}$ Js the Planck's constant, and ν the operating frequency. Other parameters in Eq. (1.15) are B the receiver bandwidth, I_d the dark current, K_B the Boltzmann's constant, T the absolute temperature, R_L the equivalent load resistance, and P_B the background noise power. When APD is used, the dark current and shot noise are increased by the multiplication process; however, the thermal noise remains unaffected. Therefore, if the photocurrent is increased by a factor of $\mathcal{M}$, avalanche multiplication factor, then the total shot noise is also increased by the same factor. The direct detection SNR for APD photodetector is given by

$$SNR = \frac{(\mathcal{M}R_0P_R)^2}{[2qB\,(R_0P_R + R_0P_B + I_{db})\,\mathcal{M}^2F + I_{ds}] + 4K_BTB/R_L} \tag{1.17}$$

In Eq. (1.17), F is the excess noise factor arising due to random nature of multiplication factor, I_{db} the bulk dark current, and I_{ds} the surface dark current.

The block diagram of direct detection receiver is given in Fig. 1.13. Since the photodetector response is insensitive to the frequency, phase, or polarization of the carrier, this type of receiver is useful only for intensity-modulated signals.

1.5.1.1 Baseband Modulation

In baseband modulation, the information signal directly modulates the LED/laser drive current and hence the optical carrier. This signal, often called baseband-modulated signal, is then transmitted through the atmospheric channel. At the receiver side, the information is recovered from the baseband-modulated signal using direct detection technique. The modulation schemes that come under this category include on-off keying (OOK) and digital pulse-position modulation (PPM). Other pulse modulation schemes like digital pulse interval modulation (DPIM), pulse amplitude and position modulation (PAPM), and differential amplitude pulse interval modulation (DAPIM) are upcoming modulation schemes, but they have not yet achieved enough popularity as compared to OOK and PPM. Most of the work on FSO systems has been carried out using OOK modulation scheme because it is simple and easy to implement [44].

In OOK, the transmission of binary data is represented by the presence or absence of light pulse, i.e., if the information bit is 1, laser is turned on for the duration T_b, and if it is 0, nothing is transmitted. OOK system requires adaptive threshold in order to deal with the fluctuating attenuation of the atmosphere. In OOK with nonreturn-to-zero (NRZ-OOK) signaling, the bit one is simply represented by an optical pulse that occupies the entire bit duration while the bit zero is represented by the absence of an optical pulse. In OOK with return-to-zero (RZ-OOK) signaling, a one bit is represented by the presence of an optical pulse that occupies part of the bit duration.

For long-distance communication, $\mathbb{M}$-PPM scheme is most widely used because of its high peak-to-average power ratio that improves its power efficiency. Also, unlike OOK it does not require adaptive threshold. In $\mathbb{M}$-PPM scheme, each symbol period is divided into $\mathbb{M}$ time slots each of duration T_s seconds, and the information is placed in one of the $\mathbb{M}$ time slots to represent a data word. Here, $\mathbb{M} = 2^n$ where n is the number of information bits. Therefore, each PPM symbol is mapped directly to n bit sequence and thus allows $\log_2\mathbb{M}$ bits within each PPM symbol. The bit-to-symbol mapping can be viewed as one-to-one assignment of symbols to each of n consecutive information bits. The PPM scheme is preferred for long-distance communication except when the transmitter power is peak power limited or system is bandwidth limited. Figures 1.14 and 1.15 show OOK and 8-PPM scheme, respectively, for the transmission of random bit sequence say 110010. It is clear that PPM scheme requires more bandwidth ($= {}^1/_{T_s}$ where $T_s = {}^{T_b \log_2 \mathbb{M}}/_{\mathbb{M}}$) than the OOK-modulated signal bandwidth ($= {}^1/_{T_b}$) and complex transceiver design due to tight synchronization requirements.

1.5.1.2 Statistical Model for Direct Detection

In case of direct detection receiver, the energy of the received signal is considered but not its phase. For every transmitted message, say X, a number of photons fall on the detector. The absorption of these photons results in an output Y from the detector that can be passed on to the demodulator and the decoder. Several direct detection statistical models are mentioned below which define the conditional probability density function of the output Y given X.

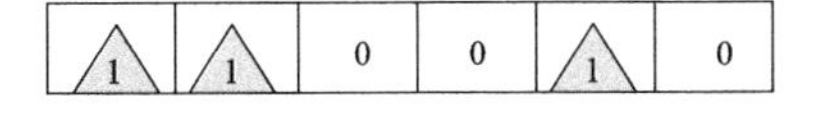

Fig. 1.14 OOK modulation scheme for the transmission of message 110010

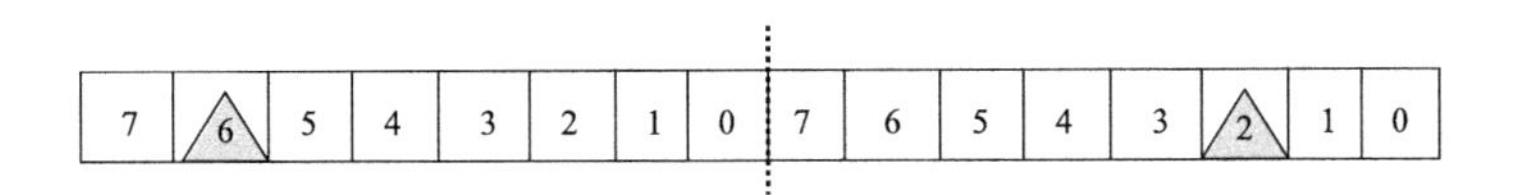

Fig. 1.15 8-PPM scheme with eight slots for the transmission of message 110010

(i) **The Poisson channel model for ideal photodetectors**
The Poisson distribution in the detection of photons [43] can be expressed as

$$f_{Y/X}(k/0) = \frac{K_0^k \exp(-K_0)}{k!} \quad k = 0, 1, 2, \ldots$$
$$f_{Y/X}(k/1) = \frac{K_1^k \exp(-K_1)}{k!} \quad k = 0, 1, 2, \ldots, \tag{1.18}$$

where K_0 and K_1 are the average number of photons detected when $X = 0$ and $X = 1$, respectively.

(ii) **The McIntyre-Conradi model for APDs**
The average number of photons absorbed over the active surface of an APD with optical power $P(t)$ in T seconds [43] can be expressed as

$$K = \frac{\eta}{h\nu} \int_0^T P(t)\, dt, \tag{1.19}$$

where h is the Planck's constant, ν the optical frequency, and η the quantum efficiency of the photodetector, defined as the ratio of average number of photons absorbed by the APD (each absorbed photon produces an electron-hole pair) to the average number of incident photons. The actual number of absorbed photons is a Poisson-distributed random variable with mean K (where $K = K_0 = K_b$ for bit "0," i.e., the sum of background photon count, or $K = K_1 = K_s + K_b$ for bit "1," i.e., the sum of actual and background photon count). In an APD, the density of the output electrical signal in response to the absorbed photons is modeled accurately by McIntyre-Conradi distribution [45]. The conditional probability density of obtaining k' electrons in response to n' absorbed photons is given by

$$f_{Y/N}(k'/n') = \frac{n'\Gamma\left(\frac{k'}{1-k_{\text{eff}}}+1\right)}{k'(k'-n')!\Gamma\left(\frac{k_{\text{eff}}k'}{1-k_{\text{eff}}}+n'+1\right)} \cdot \left[\frac{1+k_{\text{eff}}(\mathcal{M}-1)}{\mathcal{M}}\right]^{\frac{n'+k_{\text{eff}}k'}{1-k_{\text{eff}}}} \cdot \left[\frac{(1-k_{\text{eff}})(\mathcal{M}-1)}{\mathcal{M}}\right]^{k'-n'}, \tag{1.20}$$

where $\mathcal{M}$ is the average gain of the APD and k_{eff} the ionization ratio ranging from $0 < k_{\text{eff}} < 1$. Taking an average of Eq. (1.20) over the number of absorbed photons n' gives

$$f_Y(k') = \sum_{n=1}^{k'} f_{Y/N}(k'/n') \frac{K^{n'}}{n'!} \exp(-K), \qquad k' \geq 1. \tag{1.21}$$

It should be noted that the limit of summation is up to k' instead of infinity as the number of absorbed photons can never be more than the released electrons. Therefore, for $k' \in \mathbb{N}$ where $\mathbb{N}$ is the set of natural numbers, the conditional probability density function of received photons is given by [45, 46]

$$f_{Y/X}(k'/x) = \sum_{n'=1}^{k'} \frac{n'\Gamma\left(\frac{k'}{1-k_{\text{eff}}}+1\right)\left[\frac{1+k_{\text{eff}}(\mathcal{M}-1)}{\mathcal{M}}\right]^{\frac{n'+k_{\text{eff}}k'}{1-k_{\text{eff}}}} K_x^{n'} \exp(-K_x)}{k'(k'-n')!\Gamma\left(\frac{k_{\text{eff}}k'}{1-k_{\text{eff}}}|n'|\right)n'!} \cdot \left[\frac{(1-k_{\text{eff}})(\mathcal{M}-1)}{\mathcal{M}}\right]^{k-n}, \tag{1.22}$$

where $K_0 = K_b$ is the average number of photons detected when $x = 0$ and $K_1 = K_s + K_b$ is the average number of photons detected when $x = 1$. Both bulk and surface dark currents in APD will add to the background noise photon count K_b.

(iii) **Additive white Gaussian noise approximation**
The additive white Gaussian noise model is commonly used for direct detection receiver. In this case, the conditional probability density function of received photons is given by

$$f_{Y/X}(y/x) = \frac{1}{\sqrt{2\pi\sigma_x^2}} \exp\left[-(y-\mu_x)^2/2\sigma_x^2\right], \tag{1.23}$$

where $x \in \{0, 1\}$. The parameters μ_x and σ_x^2 are the mean and variance, respectively, when $X = x$. This model is often used with APD and the mean and variance in Eq. (1.23) are given as [47]

$$\begin{aligned}
\mu_0 &= \mathcal{M}K_b + I_sT_s/q, \\
\mu_1 &= \mathcal{M}(K_s + K_b) + I_sT_s/q, \\
\sigma_0^2 &= \left[\mathcal{M}^2FK_b + \frac{I_sT_s}{q} + \frac{2K_BTT_s}{q^2R_L}\right]2BT_s, \\
\sigma_1^2 &= \left[\mathcal{M}^2F(K_s + K_b) + \frac{I_sT_s}{q} + \frac{2K_BTT_s}{q^2R_L}\right]2BT_s.
\end{aligned} \tag{1.24}$$

In the above equations, $\mathcal{M}$ is the average APD gain, F the excess noise factor of APD, I_s the surface dark current, T_s the slot width, $B\,(= 1/2T_s)$ the electrical bandwidth of receiver, and K_s and K_b the average number of signal and background noise photons, respectively.

1.5.1.3 Subcarrier Modulation

In a subcarrier intensity modulation (SIM) scheme [48], the radio-frequency (RF) electrical subcarrier signal is pre-modulated with the information signal. The electrical subcarrier can be modulated using any modulation scheme like binary phase-shift keying (BPSK), quadrature phase-shift keying (QPSK), quadrature amplitude modulation (QAM), amplitude modulation (AM), frequency modulation (FM), etc. This pre-modulated signal is used to modulate the intensity of the optical carrier. At the receiver, the signal is recovered using direct detection as in IM/DD system. It does not require adaptive threshold unlike OOK scheme, and it is more bandwidth efficient than PPM scheme. Optical SIM inherits the benefits from more mature RF system; therefore, it makes the implementation process simpler. The SIM technique allows simultaneous transmission of several information signals over the optical link. The subcarrier multiplexing can be achieved by combining different modulated electrical subcarrier signals using frequency-division multiplexing (FDM), which is then used to modulate the intensity of a continuous laser source that serves as the optical carrier. Figure 1.16 shows the principle of SIM optical system for FSO link. The disadvantage of this multiplexing scheme is tight synchronization and design complexity at the receiver side.

Both baseband and SIM signals can be demodulated using direct detection/non-coherent detection techniques which are low cost, less complex, and widely used in FSO communication system. Direct detection technique can also be used with analog modulation of optical carrier. However, it is not widely used as it puts linearity constraints on the laser source and modulation technique that are difficult to achieve with present state of art. The most commonly used modulation schemes in FSO systems are shown in Fig. 1.17. The choice of a suitable modulation scheme

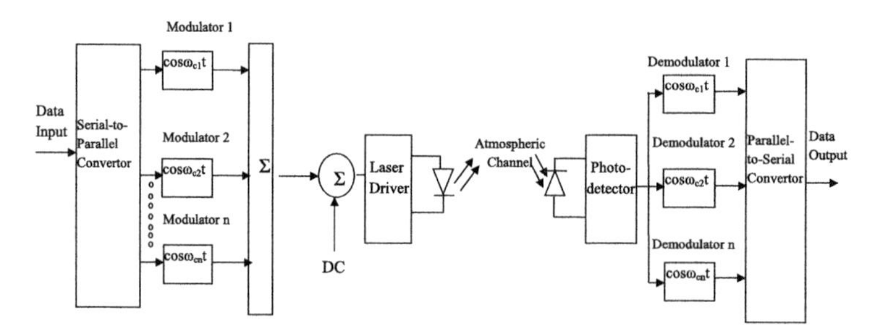

Fig. 1.16 Block diagram of SIM for FSO link

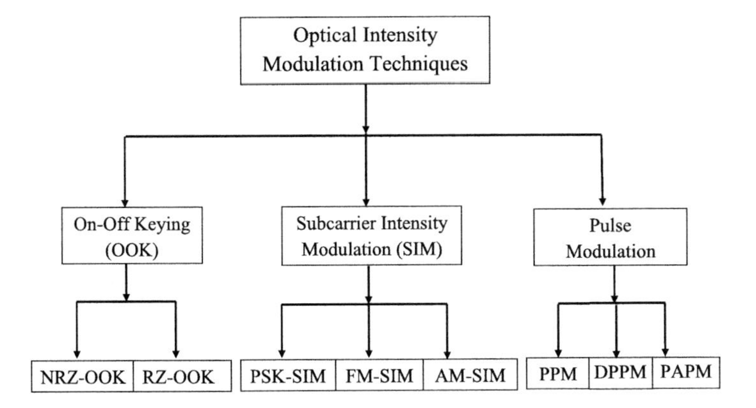

Fig. 1.17 Modulation schemes in FSO system

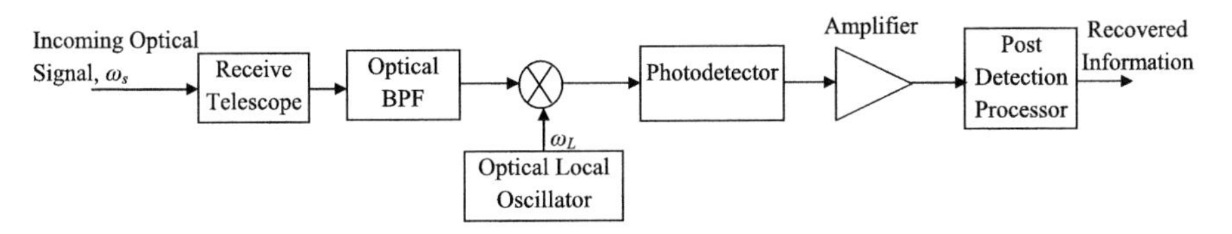

Fig. 1.18 Block diagram of coherent optical communication system

requires the trade-off between power efficiency, bandwidth requirement for a given data rate, and implementation complexity.

1.5.2 *Coherent Detection*

In coherent detection receiver, incoming signal is mixed with a locally generated coherent carrier signal from a local oscillator (LO). This mixing of incoming weak optical signal and strong LO signal at the photodetector provides signal amplification and converts the optical signal into electrical signal. The strong field of LO raises the signal level well above the noise level of the electronics circuit. Thus, the sensitivity of the coherent receiver is limited by the shot noise of the LO signal. Furthermore, because of the spatial mixing process, the coherent receiver is sensitive to signal and background noise only that falls within the same spatial temporal mode of the LO. This allows coherent detection optical receiver to operate in very strong background noise without significant degradation in the performance. The basic block diagram of coherent receiver is shown in Fig. 1.18.

Depending upon the frequency of the local oscillator ω_L and the frequency of the incoming signal ω_S, coherent detection can be categorized as heterodyne detection or homodyne detection. If ω_L is offset from ω_s by an intermediate frequency

(IF) ω_{IF}, then it is called heterodyne detection, i.e., $\omega_L = \omega_S + \omega_{IF}$. In case of homodyne detection, there is no offset between ω_L and ω_S, i.e., $\omega_{IF} = 0$ implying $\omega_L = \omega_S$. In both heterodyne and homodyne receivers, the photodetector current I_p is proportional to the optical intensity and is given as

$$I_p \propto (e_R + e_L)^2 , \tag{1.25}$$

where e_R and e_L are incoming received signal and local oscillator electric fields, respectively. The above equation can, therefore, be written as

$$I_p \propto [E_R \cos(\omega_S t + \phi_S) + E_L \cos(\omega_L t + \phi_L)]^2 , \tag{1.26}$$

where E_R and E_L are the peak incoming received and LO signals, respectively, and ϕ_s and ϕ_L are the phase of transmitted and LO signals, respectively. Solving Eq. (1.26) and removing higher frequency terms which are beyond the detector response give

$$I_p \propto \frac{1}{2}E_R^2 + \frac{1}{2}E_L^2 + 2E_R E_L \cos(\omega_L t - \omega_S t + \phi) , \tag{1.27}$$

where $\phi = \phi_S - \phi_L$. Since the signal power is proportional to the square of the electrical field, the above equation can be written as

$$I_p \propto P_R + P_L + 2\sqrt{P_R P_L} \cos(\omega_L t - \omega_S t + \phi) . \tag{1.28}$$

In the above equation, P_R and P_L are the optical power levels of incoming signal and local oscillator signal, respectively. Photocurrent in relation to incident power P_R is governed by $I_p = \eta q P_R / h\nu$. Hence, the above equation can be written as

$$I_p = \frac{\eta q}{h\nu}\left[P_R + P_L + 2\sqrt{P_R P_L} \cos(\omega_L t - \omega_S t + \phi)\right] , \tag{1.29}$$

where η is the quantum efficiency of photodetector, h the Plank' s constant, and ν the optical frequency. Generally, the local oscillator signal power is much higher than the incoming signal power, and therefore, the first term in the above equation can be neglected. After that the signal component of the photodetector current is given as

$$I_p = \frac{\eta q}{h\nu}\left[2\sqrt{P_R P_L} \cos(\omega_L t - \omega_S t + \phi)\right] . \tag{1.30}$$

For heterodyne detection, $\omega_s \neq \omega_L$ and therefore the above equation can be written as

$$I_p = \frac{\eta q}{h\nu}\left[2\sqrt{P_R P_L} \cos(\omega_{IF} t + \phi)\right] . \tag{1.31}$$

It is clear from this equation that photodetector current is centered around an IF. This IF is stabilized by incorporating the local oscillator laser in a frequency control loop. In case of homodyne detection, $\omega_L = \omega_s$ and therefore Eq. (1.31) reduces to

$$I_p = \frac{2\eta q}{h\nu}\sqrt{P_R P_L}\cos\phi = 2R_0\sqrt{P_R P_L}\cos\phi. \tag{1.32}$$

In this case, output from the photodetector is in the baseband form, and local oscillator laser needs to be phase locked to the incoming optical signal. It is clear that the signal photocurrent in both homodyne and heterodyne receivers is effectively amplified by a factor $2\sqrt{P_R P_L}$. This amplification factor has the effect of increasing the incoming optical signal level without affecting the preamplifier noise or photodetector dark current noise. This makes coherent receiver to provide higher receiver sensitivity.

Various noise contributors in coherent detection are signal shot noise, background shot noise, LO shot noise, signal-background beat noise, LO-background beat noise, background-background beat noise, and thermal noise. When the local oscillator signal power is much greater than the incoming signal power, then the dominant source of noise is due to local oscillator shot noise, and its mean square noise power is given by

$$\bar{I}_L^2 = \begin{cases} 2qR_0P_LB & \text{for PIN} \\ 2qR_0P_LB\mathcal{M}^2F & \text{for APD.} \end{cases} \tag{1.33}$$

The SNR in this case (assuming no phase difference between source and local oscillator signal, i.e., $\phi = 0$) is given by

$$SNR = \frac{I_P^2}{2qR_0P_LBF} = \frac{2R_0P_R}{qBF}. \tag{1.34}$$

The value of F is unity in case of PIN photodetector. It has been seen that coherent detection system provides larger link margin (approx 7–10 dB) over direct detection system. Coherent system can employ any modulation scheme like OOK, FSK, PSK, PPM, etc. Due to complexity and high cost of coherent receiver design, it is rarely used in FSO systems. It becomes cost-effective at high data rates and may find applications in the future.

1.5.3 Optical Orthogonal Frequency-Division Multiplexing

Optical orthogonal frequency-division multiplexing (OFDM) [49] belongs to the category of multi-carrier modulation (MCM) where the data information is carried

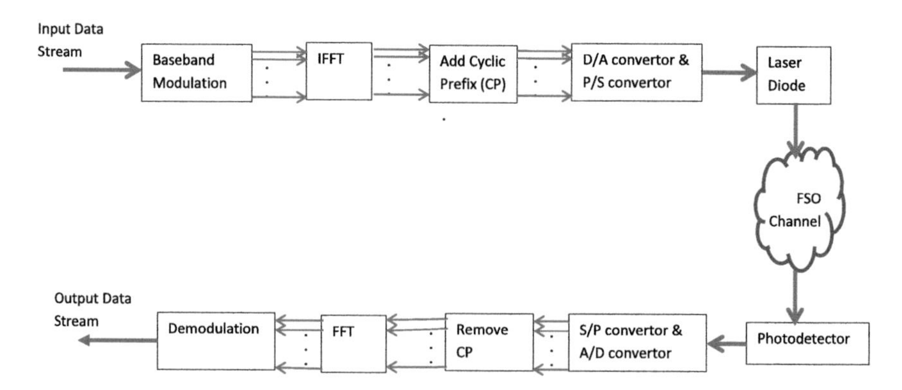

Fig. 1.19 Block diagram of OFDM based FSO system

over many lower rate subcarriers. OFDM when implemented with wireless optical system gives a very cost-effective solution for improving its performance. OFDM allows the high data rate to be divided into multiple low data rates, and they are transmitted in parallel form. The main objective of using this MCM scheme is to lower the symbol rate and provide high tolerance to deep fades that deteriorate the performance of FSO communication system. OFDM-based FSO system will exploit the advantages of both OFDM and FSO to become a good candidate for "last mile" solution for broadband connectivity. Such systems are capable of providing high spectral efficiency and increased robustness against intensity fluctuations caused by turbulent atmosphere. OFDM-based FSO system can also be used in some coding techniques. Low-density parity check (LDPC)-coded OFDM is able to give performance much better than LDPC-coded OOK over fluctuating atmospheric environment both in terms of coding gain and spectral efficiency. However, due to sensitivity of OFDM scheme to phase noise and relatively large peak-to-average power ratio, the design of OFDM for FSO system has to be implemented very carefully. Figure 1.19 shows the block diagram of OFDM-based FSO system.

In OFDM system, the input from the source is baseband modulated using any of the modulation scheme like PSK, QAM, etc. It is also known as mapping and the mapped signal is converted from serial to parallel form. This will allow high data rate stream to split into multiple low data rate narrowband subcarriers. These narrowband subcarriers experience lesser distortion than high data rates and require no equalization. Inverse fast Fourier transform (IFFT) and cyclic prefix (CP) operations are performed on the low data rate narrowband subcarrier to generate OFDM signals. It is then followed by digital-to-analog and parallel-to-serial convertors. This OFDM signal now modulates the laser diode, and it is then allowed to propagate through FSO channel. At the receiver side, after the signal is being detected by photodetector, the reverse process is carried out to recover the information signal. Since OFDM uses FFT algorithms for modulation and demodulation, such system does not require equalization. OFDM employed in

Table 1.4 Comparison of RF and optical OFDM systems

Type	Mathematical model	Speed
Wireless OFDM	Time domain multiple discrete Rayleigh fading	Can be fast for mobile environment
Optical OFDM	Continuous frequency domain dispersion	Medium

optical system is a little bit different as compared to RF-OFDM system. Table 1.4 gives the comparison of RF and optical OFDM systems.

In order to improve the power efficiency of OFDM-based FSO, three variants of OFDM schemes are used:

(i) Biased-OFDM single side-band scheme: This scheme is based on intensity modulation and is also known as "Biased-OFDM" (B-OFDM) scheme. In this case, the transmitted signal is given by

$$S(t) = S_{OFDM}(t) + D, \tag{1.35}$$

where D is the bias component. Since IM/DD does not support bipolar signals, the bias component D has to be sufficiently large so that when it is added to $S_{OFDM}(t)$, it results in a nonnegative component. The main disadvantage of B-OFDM scheme is the poor efficiency.

(ii) Clipped-OFDM single side-band scheme: It is based on single side-band transmission with clipping of the negative portion of OFDM signal after bias addition. The conversion of DSB to SSB can be made in two ways: (a) by the use of Hilbert transformation of inphase signal as the quadrature signal in the electrical domain or (b) by the use of optical filter. By choosing the optimum bias value, the power efficiency of clipped OFDM (C-OFDM) can be improved as compared to B-OFDM.

(iii) Unclipped-OFDM single side-band scheme: This scheme employs $LiNbO_3$ Mach-Zehnder modulator (MZM) to improve its power efficiency. To avoid distortion due to clipping, the information signal is transmitted by modulating the electric field so that the negative part of OFDM is given to the photodetector. The distortion introduced by the photodetector is removed by proper filtering and in recovered signal distortion is insignificant. Unclipped OFDM (U-OFDM) is less power efficient than C-OFDM, but is significantly more power efficient than B-OFDM.

The optical OFDM signal at the receiver can be detected by using non-coherent or coherent technique. The main features of non-coherent and coherent OFDM systems are as follows:

(i) Non-coherent OFDM: It uses direct detection at the receiver but it can further be classified into two categories according to how the OFDM signal is generated: (a) linearly mapped direct detection optical OFDM (DDO-OFDM or

LM-DDO-OFDM) where the optical OFDM spectrum is a replica of baseband OFDM and (b) nonlinearly mapped DDO-OFDM (or NLM-DDO-OFDM) where the optical OFDM spectrum does not display a replica of baseband OFDM.

(ii) Coherent OFDM: It achieves high spectral efficiency by overlapping subcarrier spectrum while avoiding interference with the help of coherent detection and signal set orthogonality. Coherent OFDM improves the sensitivity of the receiver and increases the robustness against polarization dispersion. The synergies between coherent optical communication and OFDM are twofold. OFDM brings computation efficiency of coherent systems along with ease of channel and phase estimation. Also, it brings linearity in RF to optical (RTO) up conversion and optical to RF (OTR) down conversion. However, it increases the cost and design complexity of the system.

1.6 Eye Safety and Regulations

While designing an FSO link, the designer has to ensure that the chosen operating wavelength has to be eye and skin safe. This means that the laser should not pose any kind of danger to the people who might encounter the communication beam. Figure 1.20 clearly shows the region where different wavelengths of light get absorbed in the human eye.

Microwaves and gamma rays are absorbed by the human eye and can cause high degree of damage to lenses and retina. Near-ultraviolet (UV) wavelengths are absorbed in the lens making them cloudy (cataract) which leads to dim vision or

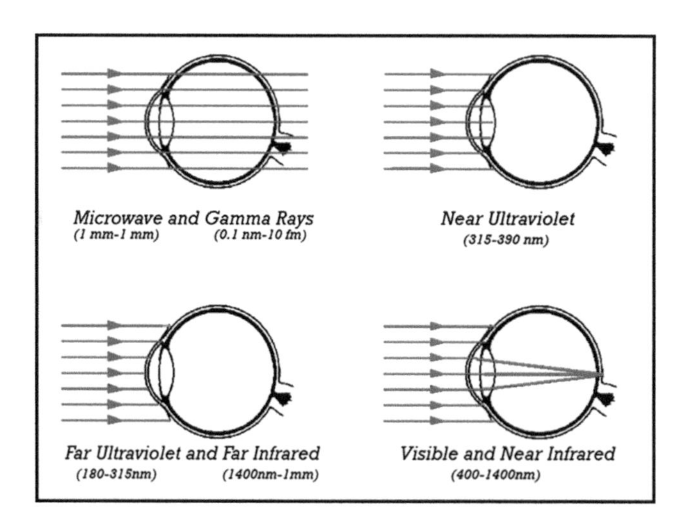

Fig. 1.20 Pictorial representation of light absorption in the eye for different wavelengths [50]

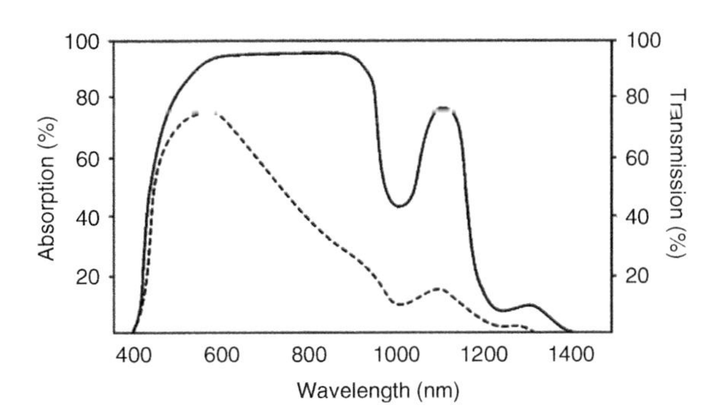

Fig. 1.21 Absorption of light vs. wavelength

blurring. In far UV and IR regions, wavelengths are absorbed in the cornea and produce an effect called photokeratitis which can lead to pain/watering in the eye or pigmentation in the cornea. Visible and near IR region (wavelengths used in FSO communication) has the highest potential to damage the retina of the eye, and it can lead to permanent loss of vision that cannot be healed by any surgery. The range of wavelengths between 400 and 1400 nm can cause potential eye hazards or even skin damage [25]. The impact of laser injury is more significant in case of eyes than the skin as the outer layer of the eye, i.e., the cornea, acts as a band-pass filter to the wavelengths. Therefore, the cornea will be transparent to these wavelengths, and the energy emitted by the light sources will get focused on the retina and may cause damage to the eye due to increase in concentration of the optical energy. However, the light below 400 nm and above 1400 nm is absorbed by the cornea and does not reach the retina. The absorption coefficient of the cornea is more for higher wavelengths than (>1400 nm) than for the shorter wavelengths as can be observed from the Fig. 1.21.

Lasers can cause damage to our skin by causing thermal burns or photochemical reaction. The penetration of laser beam inside the human body depends upon the choice of operating wavelength. UV rays are absorbed by the outer layer of the skin and cause skin cancer or premature aging of the skin. Exposure of high-intensity beam for very long period can cause thermal burns or skin rashes. IR radiation can penetrate deep into skin leading to thermal burns. Therefore, it is very essential to regulate the laser power to ensure the safety of the human eye and skin.

Various international standard bodies (such as American National Standards Institute (ANSI) Z136 in the USA, Australian/New Zealand (AS/NZ) 2211 Standard in Australia, and International Electrotechnical Commission (IEC) 60825 internationally) provide safety guidelines of laser beam depending upon their wavelength and power. Laser Institute of America (LIA) is an organization that promotes the safe use of lasers and provides laser safety information. These standards are global benchmark for laser safety, and they are used as guidelines for various manufacturers. Every organization has its own way of laser classification, and accordingly, safety precautions and administrative control measures have to be taken. The classification of the laser is based on whether or not the maximum permissible emission (MPE) is longer or shorter than the human aversion response.

MPE is a quantity that specifies a certain level up to which the unprotected human eye can be exposed to laser beam without any hazardous effect to the eye or skin. Aversion response is the autonomic response (within 0.25 s) of the blinking eye when it moves away from the bright source of light. Another quantity that determines classification of lasers is accessible emission limit (AEL) which is the mathematical product of two terms, i.e., MPE limit and limiting area (LA) factor. Therefore, $AEL = MPE \times LA$.

Based upon MPE and AEL calculations, lasers are broadly classified into four groups, i.e., Class 1 through Class 4. The lower classifications (Class 1 and 2) have minimum power and therefore do not require protective eye wear. This class has extended MPE measurements as the human eye will avert from the bright light long before the beam can injure the unprotected eye. The higher classification (Class 3R, 3B, and 4) has high power levels; therefore, proper eye safety precautions have to be taken during laser operations. In this class MPE is shorter than aversion response. Table 1.5 gives the comparison of laser classification according to IEC and ANSI standards.

Table 1.6 presents the AEL for two most commonly used wavelengths in FSO communication systems. It is evident from the table that for Class 1 and 2, lasers operating at 1550 nm are almost 50 times more powerful than lasers at shorter operating wavelength, i.e., 850 nm. Also, the combination of low attenuation, high component availability, and eye safety at 1550 nm wavelength makes it a preferred choice for FSO communication. Lasers operating at 1550 nm wavelength when used with erbium-doped fiber amplifier (EDFA) technology are capable of providing high data rates (>2.5 Gbps) and high power.

Class 1 and Class 1M lasers are preferred for terrestrial FSO communication links as their radiations are safe under all circumstances. IEC60825-12 [51] covers the safety standard for free-space optical communication links, and it lists out the requirements like power, aperture size, distance, and power density for 850 and 1550 nm wavelengths which are presented in Table 1.7. Higher classes of lasers are also used for FSO communication links for long distance communication or deep space missions. To ensure the safety of these systems, they are installed on higher platforms like rooftops or towers to prevent any kind of human injury.

It is to be noted that high-powered pulsed lasers are more dangerous than lower power continuous lasers. However, lower power laser beams can also be hazardous when given long-term exposure.

1.7 Applications of FSO Communication Systems

Applications of FSO communication systems range from short range (<1 km) to long range and space applications. It provides broadband solution (high data rates without cabling) for connecting end users to the backbone. Short-range application provides last mile access by connecting various towers, buildings, etc. in urban areas

Table 1.5 Laser classification according to IEC and ANSI standards

Classification	IEC 60825	ANSI-Z136.1
Class 1	Very low power lasers and are safe under reasonably foreseeable conditions of operation. This class is exempted from all beam-hazard control measures. It includes optical instruments for intrabeam viewing.	
Class1M	Low power lasers operating between 302.5 nm and 4000 nm wavelengths and are safe under reasonably foreseeable conditions except when used with optical instruments such as collecting lens, binoculars, telescope, etc. These lasers produce either collimated beams with large beam size or highly divergent beams.	N/A
Class 2	Low power laser operating between 400 nm to 700 nm (visible range). This laser class can be continuous wave (CW) or repetitively pulsed lasers. It is safe to use if it emits energy below Class 1 AEL for emission duration of less than 0.25 sec (i.e., the time period of the human eye aversion response). It have an average radiant power of 1mW or less.	
Class 2M	Low power laser operating between 400 nm to 700 nm (visible range). It can cause optical hazards when viewed with optical instruments such as collection lens, telescope, etc. Any emissions outside this wavelength region must be below the Class 1M AEL.	N/A
Class 3R	Average power lasers operating between 302.5 nm and 10^6 nm. The accessible emission limit is within 5 times the Class 2 AEL for visible range wavelengths and within 5 times the Class 1 AEL for wavelengths outside this region. It is unsafe to view the beam directly with diameter > 7mm.	N/A
Class 3A	N/A	Average power lasers operating between 302.5 nm and 10^6 nm. The accessible emission limit is within 5 times the Class 2 AEL for visible range wavelengths and between 1 and 5 times the Class 1 AEL for wavelengths outside this region. It is unsafe to view the beam directly with diameter > 7mm.
Class 3B	Average power lasers that cannot emit an average radiant power greater than 0.5 Watts for an exposure time equal to or greater than 0.25 seconds. It is unsafe to view the beam directly but are normally safe when view diffused reflections	
Class 4	High power lasers and are very dangerous both under intrabeam and diffuse reflection viewing conditions. They may also cause skin injuries and are potential fire hazards.	

where digging of cables is a difficult task. It includes point-to-point or point-to-multipoint links or broadband links. Various applications of FSO systems are listed below:

(i) Enterprise connectivity: FSO link can easily be deployed to connect various tower/building enabling local area connectivity. It can also be extended to connect metropolitan area fiber rings, connect new networks, and provide high-speed network expansion.

Table 1.6 Accessible emission limits for 850 and 1550 nm according to IEC standard

Laser classification	Average output optical power (mW)	
	850 nm	1550 nm
1	<0.22	<10
2	Used only for 400–700 nm and has same AEL as Class 1	
3R	0.22–2.2	10–50
3B	2.2–500	50–500
4	>500	>500

Table 1.7 Various requirements of Class 1 and 1M lasers for 850 and 1550 nm [52]

Classification	Power (mW)	Aperture size (mm)	Distance (m)	Power density (mW/cm^2)
850 nm Wavelength				
Class 1	0.78	7	14	2.03
		50	2000	0.04
Class 1M	0.78	7	100	2.03
	500	7	14	1299.08
		50	2000	25.48
1550 nm Wavelength				
Class 1	10	7	14	26
		25	2000	2.04
Class 1M	10	3.5	100	103.99
	500	7	14	1299.88
		25	2000	101.91

(ii) Fiber backup: In case of optical fiber link failure, FSO link can be deployed as a backup link to ensure availability of the system.

(iii) Point-to-point links: It coves inter- (LEO-LEO) and intra-orbital (LEO-GEO) links and satellite-to-ground/ground-to-satellite link. This type of link requires good pointing and tracking system. Here, the output power of the transmitter, power consumption, size, mass, and deployment cost increase with the link range.

(iv) Point-to-multipoint links: Multi-platform multi-static sensing, interoperable satellite communications, and shared spaceborne processing are unique network application of FSO system.

(v) Hybrid wireless connection/network redundancy: FSO communication is prone to weather conditions like fog, heavy snow, etc. In order to obtain 100 % availability of the network, FSO links can be combined with microwave links that operate at high frequencies (in GHz range) and offer comparable data rates.

(vi) Disaster recovery: FSO communication system provides high-capacity scalable link in case of collapse of existing communication network.

(vii) Backhaul for cellular networks: With the advent of 3G/4G cellular communication, there is a growing challenge to increase the backhaul capacity between the cell towers to cope up with the increase in demand of broadband mobile customers. The viable backhaul options for 4G network are to deploy fiber-optic cables or to install FSO connection between towers. Deploying fiber cables is time-consuming and an expensive task. So FSO communication system plays an important role in providing backhaul capacity for cellular networks.

1.8 Summary

This chapter begins with the discussion of various types of optical wireless communication ranging from indoor IR to outdoor FSO communication. However, the chapter mainly focuses on outdoor terrestrial FSO communication link. It presents the advantages of optical carrier over RF carrier where the transmission rate can exceed 10 Gbps and can find its application in enterprise connectivity, video surveillance and monitoring, disaster recovery, backhaul for cellular systems, etc. Various technologies used in FSO communication system, i.e., direct detection, coherent detection, and OFDM, are presented in this chapter. The choice of operating wavelength based on absorption losses and component availability in market is also discussed.

Bibliography

1. R.F. Lucy, K. Lang, Optical communications experiments at 6328 Å and 10.6 μ. Appl. Opt. **7**(10), 1965–1970 (1968)
2. M.S. Lipsett, C. McIntyre, R. Liu, Space instrumentation for laser communications. IEEE J. Quantum Electron. **5**(6), 348–349 (1969)
3. I. Arruego, H. Guerrero, S. Rodriguez, J. Martinez-Oter et al., OWLS: a ten-year history in optical wireless links for intra-satellite communications. IEEE J. Sel. Areas Commun. **27**(9), 1599–1611 (2009)
4. S. Kazemlou, S. Hranilovic, S. Kumar, All-optical multihop free-space optical communication systems. J. Lightwave Technol. **29**(18), 2663–2669 (2011)
5. K. Hirabayashi, T. Yamamoto, S. Hino, Optical backplane with free-space optical interconnections using tunable beam deflectors and a mirror for bookshelf-assembled terabit per second class asynchronous transfer mode switch. Opt. Eng. **37**, 1332–1342 (2004)
6. N. Savage, Linking with light. IEEE Spectr. (2002). [Weblink: http://spectrum.ieee.org/semiconductors/optoelectronics/linking-with-light]
7. G. Forrester, Free space optics, in *Digital Air Wireless*. [Weblink: http://www.digitalairwireless.com/wireless-blog/2013-07/free-space-optics.html]
8. http://andy96877.blogspot.com/p/visible-light-communication-vlc-isdata.html. Visible light communication- VLC & Pure VLC^{TM}. [Weblink: http://andy96877.blogspot.com/p/visible-light-communication-vlc-is-data.html]
9. Weblink: http://lasercommunications.weebly.com/
10. Weblink: http://artolink.com
11. Weblink: http://www.fsona.com

12. L.C. Andrews, R.L. Phillips, *Laser Beam Propagation through Random Medium*, 2nd edn. (SPIE Optical Engineering Press, Bellinghan, 1988)
13. www.laserlink.co.uk. Technical report
14. A.M. Street, P.N. Stavrinou, D.C. O'Brien, D.J. Edward, Indoor optical wireless systems – a review. Opt. Quantum Electron. **29**, 349–378 (1997)
15. Z. Ghassemlooy, A. Hayes, Indoor optical wireless communication systems – part I: review. Technical report (2003)
16. A.P. Tang, J.M. Kahn, K.P. Ho, Wireless infrared communication links using multi-beam transmitters and imaging receivers, in *Proceedings of IEEE International Conference on Communications*, Dallas, 1996, pp. 180–186
17. J.B. Carruthers, J.M. Kahn, Angle diversity for nondirected wireless infrared communication. IEEE Trans. Commun. **48**(6), 960–969 (2000)
18. G. Yun, M. Kavehrad, Spot diffusing and fly-eye receivers for indoor infrared wireless communications, in *Proceedings of the 1992 IEEE Conference on Selected Topics in Wireless Communications*, Vancouver, 1992, pp. 286–292
19. R. Ramirez-Iniguez, R.J. Green, Indoor optical wireless communications, in *IEE Colloquium on Optical Wireless Communication*, vol. 128 (IET, 1999), pp. 14/1–14/7. [Weblink: http://ieeexplore.ieee.org/abstract/document/793885/]
20. J. Li, J.Q. Liu, D.P. Taylor, Optical communication using subcarrier PSK intensity modulation through atmospheric turbulence channels. IEEE Trans. Commun. **55**(8), 1598–1606 (2007)
21. J.H. Franz, V.K. Jain, *Optical Communications: Components and Systems* (Narosa Publishing House, Boca Raton, 2000)
22. H. Hemmati, *Deep Space Optical Communications* (John Wiley & Sons, Hoboken, 2006)
23. A. Jurado-Navas, J.M. Garrido-Balsells, J. Francisco Paris, M. Castillo-Vázquez, A. Puerta-Notario, Impact of pointing errors on the performance of generalized atmospheric optical channels. Opt. Exp. **20**(11), 12550–12562 (2012)
24. Weblink: http://www.cie.co.at/, 28 Feb 2012
25. O. Bader, C. Lui, Laser safety and the eye: hidden hazards and practical pearls. Technical report: American Academy of Dermatology, Lion Laser Skin Center, Vancouver and University of British Columbia, Vancouver, B.C., 1996
26. G.D. Fletcher, T.R. Hicks, B. Laurent, The SILEX optical interorbit link experiment. IEEE J. Electr. Commun. Eng. **3**(6), 273–279 (2002)
27. K.E. Wilson, An overview of the GOLD experiment between the ETS-VI satellite and the table mountain facility. TDA progress report 42-124, Communication Systems Research Section, pp. 8–19, 1996. [Weblink: https://ntrs.nasa.gov/search.jsp?R=19960022219]
28. T. Dreischer, M. Tuechler, T. Weigel, G. Baister, P. Regnier, X. Sembely, R. Panzeca, Integrated RF-optical TT & C for a deep space mission. Acta Astronaut. **65**(11), 1772–1782 (2009)
29. G. Baister, K. Kudielka, T. Dreischer, M. Tüchler, Results from the DOLCE (deep space optical link communications experiment) project. Proc. SPIE Free-Space Laser Commun. Technol. XXI **7199**, 71990B-1–71990B-9 (2009)
30. D.E. Smith, M.T. Zuber, H.V. Frey, J.B. Garvin, J.W. Head, D.O. Muhleman et al., Mars orbiter laser altimeter: experiment summary after first year of global mapping of Mars. J. Geophys. Res. **106**(E10), 23689–23722 (2001)
31. General Atomics Aeronautical Systems, Inc., *GA-ASI and TESAT Partner to Develop RPA-to-spacecraft Lasercom Link*, 2012. [Weblink: http://www.ga-asi.com/ga-asi-and-tesat-partner-to-develop-rpa-to-spacecraft-lasercom-link]
32. G.G. Ortiz, S. Lee, S.P. Monacos, M.W. Wright, A. Biswas, Design and development of a robust ATP subsystem for the altair UAV-to-ground lasercomm 2.5-Gbps demonstration. Proc. SPIE Free-Space Laser Commun. Technol. XV **4975**, 1–12 (2003)
33. D. Isbel, F. O'Donnell, M. Hardin, H. Lebo, S. Wolpert, S. Lendroth, Mars polar lander/deep space 2. Technical report, National Aeronautics and Space Administration, 1999
34. Y. Hu, K. Powell, M. Vaughan, C. Tepte, C. Weimer et al., Elevation Information in Tail (EIT) technique for lidar altimetry. Opt. Exp. **15**(22), 14504–14515 (2007)

35. N. Perlot, M. Knapek, D. Giggenbach, J. Horwath, M. Brechtelsbauer et al., Results of the optical downlink experiment KIODO from OICETS satellite to optical ground station oberpfaffenhofen (OGS-OP). Proc. SPIE, Free-Space Laser Commun. Technol. XIX Atmos. Prop. Electromag. Waves **6457**, 645704–1–645704–8 (2007)
36. V. Cazaubiel, G. Planche, V. Chorvalli, L. Hors, B. Roy, E. Giraud, L. Vaillon, F. Carré, E. Decourbey, LOLA: a 40,000 km optical link between an aircraft and a geostationary satellite, in *Proceedings of 6th International Conference on Space Optics*, Noordwijk, June 2006
37. R. Beer, T.A. Glavich, D.M. Rider, Tropospheric emission spectrometer for the Earth observing system's Aura satellite. Appl. Opt. **40**(15), 2356–2367 (2001)
38. K.E. Wilson, J.R. Lesh, An overview of Galileo optical experiment (GOPEX). Technical report: TDA progress report 42-114, Communication Systems Research Section, NASA, 1993
39. K. Nakamaru, K. Kondo, T. Katagi, H. Kitahara, M. Tanaka, An overview of Japan's Engineering Test Satellite VI (ETS-VI) project, in *Proceedings of IEEE, Communications, International Conference on World Prosperity Through Communications*, Boston, vol. 3, 1989, pp. 1582–1586
40. Y. Fujiwara, M. Mokuno, T. Jono, T. Yamawaki, K. Arai, M. Toyoshima, H. Kunimori, Z. Sodnik, A. Bird, B.a. Demelenne, Optical inter-orbit communications engineering test satellite (OICETS). Acta Astronaut. **61**(1–6), 163–175 (2007). Elsevier
41. K. Pribil, J. Flemmig, Solid state laser communications in space (solacos) high data rate satellite communication system verification program. Proc. SPIE, Space Instrum. Spacecr. Opt. **2210**(39), 39–49 (1994)
42. Z. Sodnik, H. Lutz, B. Furch, R. Meyer, Optical satellite communications in Europe. Proc. SPIE, Free-Space Laser Commun. Technol. XXII **7587**, 758705-1–758705-9 (2010)
43. R.M. Gagliardi, S. Karp, *Optical Communications*, 2nd edn. (John Wiley & Sons, New York, 1995)
44. X. Zhu, J.M. Kahn, Free space optical communication through atmospheric turbulence channels. IEEE Trans. Commun. **50**(8), 1293–1300 (2002)
45. R.J. McIntyre, The distribution of gains in uniformly multiplying avalanche photodiodes: theory. IEEE Trans. Electron Devices **19**(6), 703–713 (1972)
46. P.P. Webb, R.J. McIntyre, J. Conradi, Properties of avalanche photodiodes. RCA Rev. **35**, 234–278 (1974)
47. M. Srinivasan, V. Vilnrotter, Symbol-error probabilities for pulse position modulation signaling with an avalanche photodiode receiver and Gaussian thermal noise. TMO progress report 42–134, Jet Propulsion Laboratory, California Institute of Technology, Pasadena, Aug 1998
48. W.O. Popoola, Z. Ghassemlooy, BPSK subcarrier intensity modulated free-space optical communication in atmospheric turbulence. J. Lightwave Technol. **27**(8), 967–973 (2009)
49. D. Barros, S. Wilson, J. Kahn, Comparison of orthogonal frequency-division multiplexing and pulse-amplitude modulation in indoor optical wireless links. IEEE Trans. Commun. **60**(1), 153–163 (2012)
50. Weblink: http://www.chem.wwu.edu/dept/dept/tutorial/
51. Safety of laser products-part 12: safety of free space optical communication systems used for transmission of information. Technical report: IEC 60825-12, 2004
52. Weblink: http://web.mst.edu/~mobildat/Free%20Space%20Optics/

Chapter 2
Free-Space Optical Channel Models

2.1 Atmospheric Channel

The concentric layers around the surface of the Earth are broadly classified into two regions: homosphere and heterosphere. The homosphere covers the lower layers ranging from 0–90 km. Heterosphere lies above homosphere above 90 km. The homospheric region of atmosphere is composed of various gases, water vapors, pollutants, and other chemicals. Maximum concentrations of these particles are near the Earth surface in the troposphere that extends up to 20 km. Figure 2.1 depicts the broad classification of various layers of the atmosphere, and details of each layer with their temperature values are shown in Fig. 2.2.

The density of particles decreases with the altitude up through the ionosphere (region of upper atmosphere that extends from about 90 to 600 km and contains ionized electrons due to solar radiations). These ionized electrons form a radiation belt around the surface of the Earth. These atmospheric particles interact with all signals that propagate through the radiation belt and lead to deterioration of the received signal due to absorption and scattering. Absorption is the phenomenon where the signal energy is absorbed by the particles present in the atmosphere resulting in the loss of signal energy and gain of internal energy of the absorbing particle. In scattering, there is no loss of signal energy like in absorption, but the signal energy is redistributed (or scattered) in arbitrary directions. Both absorption and scattering are strongly dependent upon operating wavelength and will lead to decrease in received power level. These effects become more pronounced when the operating wavelength of the transmitted signal is comparable with the cross-sectional dimensions of the atmospheric particles. Figure 2.3 shows transmittance (or attenuation) effects as a function of wavelength using MODTRAN software package. The output of MODTRAN in clear weather conditions is plotted for wavelength up to 3 μm. It is clear from this figure that peaks in attenuation at specific wavelengths is due to absorption by atmospheric particles and therefore the choice

H. Kaushal et al., *Free Space Optical Communication*, Optical Networks,
DOI 10.1007/978-81-322-3691-7_2

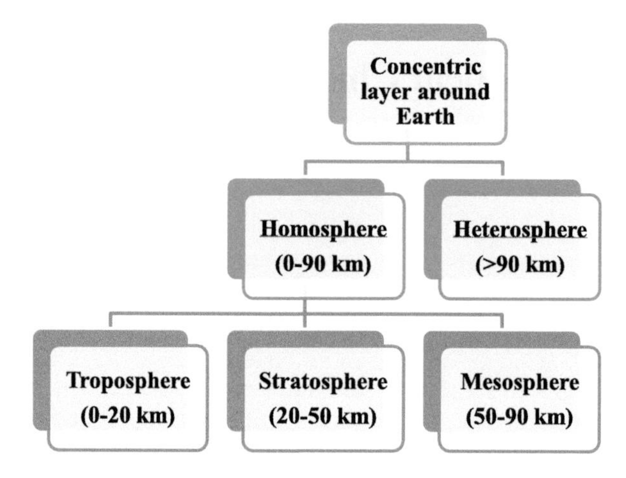

Fig. 2.1 Broad classification of atmospheric layers

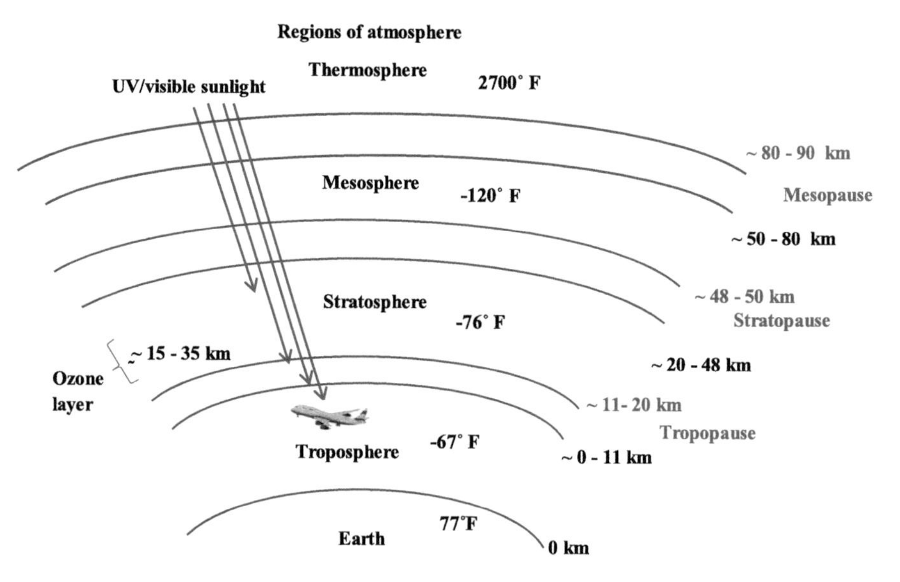

Fig. 2.2 Various atmospheric layers with corresponding temperatures

of wavelength has to be done very wisely in the high transmissive band for FSO communication links.

The atmospheric condition in FSO channel can be broadly classified into three categories, namely, clear weather, clouds, and rain. Clear weather conditions are characterized by long visibility and relatively low attenuation. Cloudy weather conditions range from mist or fog to heavy clouds and are characterized by low visibility, high humidity, and large attenuation. Rain is characterized by the presence of rain droplets of variable sizes, and it can produce severe effects depending upon rainfall rate.

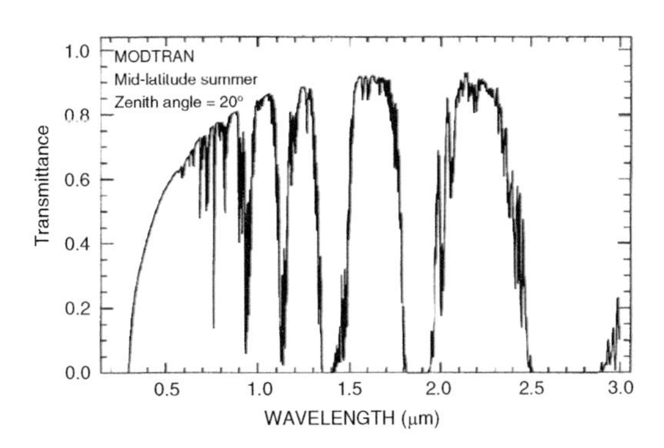

Fig. 2.3 Atmospheric transmittance (attenuation) vs. wavelength [1] (Disclaimer: This image is from a book chapter that was produced by personnel of the US Government; therefore it cannot be copyrighted and is in the public domain)

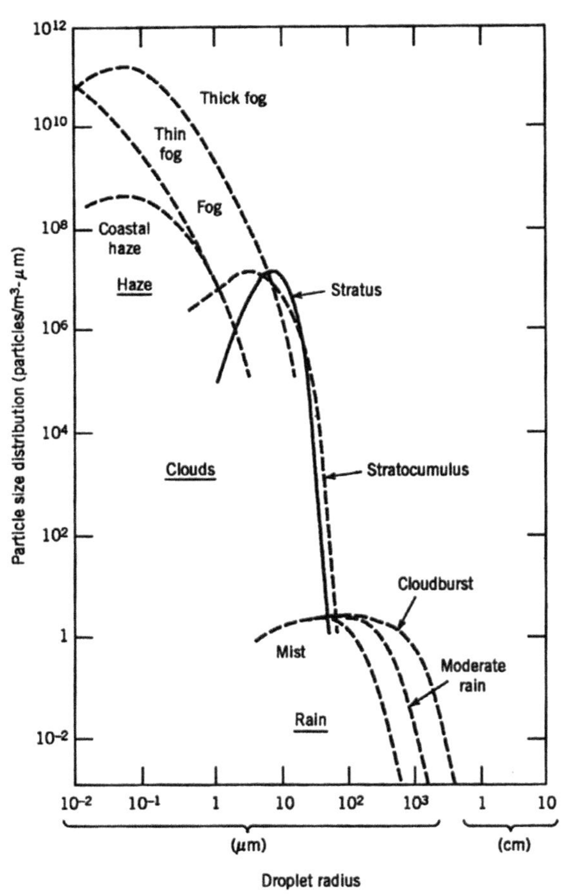

Fig. 2.4 Average particle size and corresponding particle density in atmosphere [2]

Various atmospheric conditions can be represented by size of the particle (i.e., cross-sectional dimension relative to operating wavelength) and the particle density (i.e., volumetric concentration of the particles). Figure 2.4 shows the average droplet size and its distribution for various cloudy and rainy conditions. It is seen that conditions may vary from high density and small particle size like in the case of mist and fog to low density and large particle size during heavy rain. It should be noted that Fig. 2.4 gives average parameters, whereas real atmospheric conditions may undergo various temporal changes.

2.1.1 *Atmospheric Losses*

The atmospheric channel consists of various gases and other tiny particles like aerosols, dust, smoke, etc., suspended in the atmosphere. Besides these, large precipitation due to rain, haze, snow, and fog is also present in the atmosphere. Each of these atmospheric constituents results in the reduction of the power level, i.e., attenuation of optical signal due to several factors, including absorption of light by gas molecules, Rayleigh, or Mie scattering. Various types of losses encountered by the optical beam when propagating through the atmospheric optical channel are described in this section.

In FSO communication system, when the optical signal propagates through the atmosphere, it experiences power loss due to several factors as discussed in the following sections.

2.1.1.1 Absorption and Scattering Losses

The loss in the atmospheric channel is mainly due to absorption and scattering processes. At visible and IR wavelengths, the principal atmospheric absorbers are the molecules of water, carbon dioxide, and ozone [3, 4]. The attenuation experienced by the optical signal when it passes through the atmosphere can be quantified in terms of optical depth τ which correlates with power at the receiver P_R and the transmitted power P_T [5] as

$$P_R = P_T \exp\left(-\tau\right). \tag{2.1}$$

The ratio of power received to the power transmitted in the optical link is called atmospheric transmittance T_a $(= P_R/P_T)$.

When the optical signal propagates at a zenith angle θ, the transmittance factor is then given by $T_\theta = T_a \sec\left(\theta\right)$. The atmospheric transmittance T_a and the optical depth τ are related to the atmospheric attenuation coefficient γ and the transmission range R as follows:

$$T_a = \exp\left(-\int_0^R \gamma\left(\rho\right) d\rho\right) \tag{2.2}$$

and

$$\tau = \int_0^R \gamma(\rho)\, d\rho. \tag{2.3}$$

In both the cases, the loss in dB that the beam experiences during propagation through the atmosphere can be calculated using the following equation

$$Loss_{prop} = -10 \log_{10} T_a. \tag{2.4}$$

In the first case, this loss in dB will be 4.34τ. Hence, an optical depth of 0.7 gives a loss of 3 dB.

The attenuation coefficient is the sum of the absorption and scattering coefficients from aerosols and molecular constituents of the atmosphere and is given by [6]

$$\gamma(\lambda) = \underbrace{\alpha_m(\lambda)}_{\substack{\text{Molecular}\\ \text{absorb. coeff.}}} + \underbrace{\alpha_a(\lambda)}_{\substack{\text{Aerosol}\\ \text{absorb. coeff.}}} + \underbrace{\beta_m(\lambda)}_{\substack{\text{Molecular}\\ \text{scatt. coeff.}}} + \underbrace{\beta_a(\lambda)}_{\substack{\text{Aerosol}\\ \text{scatt. coeff.}}}. \tag{2.5}$$

The first two terms in the above equation represent the molecular and aerosol absorption coefficients, respectively, while the last two terms are the molecular and aerosol scattering coefficients, respectively. The atmospheric absorption is a wavelength-dependent phenomenon. Some typical values of molecular absorption coefficients are given in Table 2.1 for clear weather conditions. The wavelength range of FSO communication system is chosen to have minimal absorption. This is referred to as atmospheric transmission windows. In this window, the attenuation due to molecular or aerosol absorption is less than 0.2 dB/km. There are several transmission windows within the range of 700–1600 nm. Majority of FSO systems are designed to operate in the windows of 780–850 and 1520–1600 nm.

The scattering process results in the angular redistribution of the optical energy with and without wavelength change. It depends upon the radius r of the particles encountered during the propagation process. If $r < \lambda$, the scattering process is classified as Rayleigh scattering; if $r \approx \lambda$, it is Mie scattering. For $r > \lambda$, the scattering process can be explained using the diffraction theory (geometric optics). The scattering process due to various scattering particles present in the atmosphere channel is summarized in Table 2.2. Out of various scattering particles like air

Table 2.1 Molecular absorption at typical wavelengths [7]

S.No	Wavelength (nm)	Molecular absorption (dB/km)
1.	550	0.13
2.	690	0.01
3.	850	0.41
4.	1550	0.01

Table 2.2 Size of various atmospheric particles present in the optical channel and type of scattering process

Type	Radius (μm)	Scattering process
Air molecules	0.0001	Rayleigh
Haze particle	0.01–1	Rayleigh-Mie
Fog droplet	1–20	Mie-Geometrical
Rain	100–10,000	Geometrical
Snow	1000–5000	Geometrical
Hail	5000–50,000	Geometrical

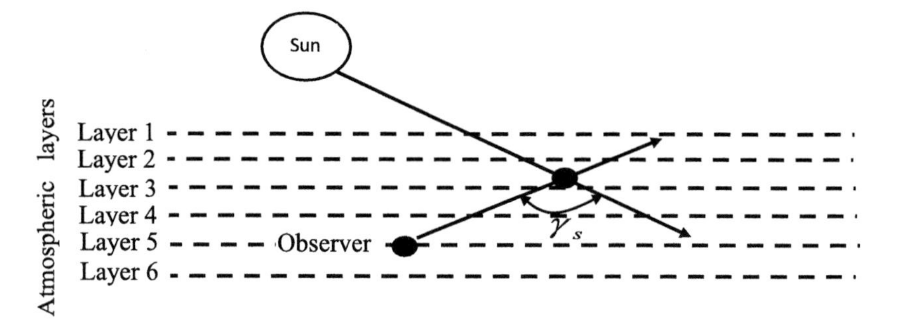

Fig. 2.5 Sky radiance due to scattering mechanism

molecules, haze particles, fog droplets, snow, rain, hail, etc., the wavelength of fog particles is comparable with the wavelength of FSO communication system. Therefore, it plays a major role in the attenuation of an optical signal.

Atmospheric scattering not only attenuates the signal beam in the atmosphere, but it is also the primary cause for sky radiance which introduces noise in daytime communication [8]. Sky radiance is due to the scattering of solar photons along the atmospheric propagating path, and it gives rise to unwanted background noise which degrades the signal-to-noise ratio at the receiver. The received background noise depends upon the geometry of the receiver and relative location of the Sun and the transmitter. Figure 2.5 shows the scattering mechanism for the layered model of the atmosphere. The atmosphere is considered to be modeled as multiple layers with each layer consisting of homogeneous mixture of gases and aerosols. The scattering angle γ_s is the angle formed between the forward direction of the Sun radiation and the point of observation. It is seen that higher the concentration of the scatterers, more will be the sky radiance. As the angular distance between the observation direction and the Sun decreases, there is an increase in sky radiance. Within 30° from the Sun, sky radiance is greatly dominated by aerosol contribution. As the angular distance from the Sun increases, the dominant source of background radiation is due to Rayleigh scattering.

2.1.1.2 Free-Space Loss

In an FSO communication system, the largest loss is usually due to "space loss," i.e., the loss in the signal strength while propagating through free space. The space loss factor is given by

$$L_s = \left(\frac{\lambda}{4\pi R}\right)^2, \tag{2.6}$$

where R is the link range. Due to dependence on wavelength, the free-space loss incurred by an optical system is much larger (i.e., the factor L_s is much smaller) than in an RF system. Besides the space loss, there are additional propagation losses if the signal passes through a lossy medium, e.g., a planetary atmosphere. Many optical links like deep space optical links do not have additional space loss as they do not involve the atmosphere.

2.1.1.3 Beam Divergence Loss

As the optical beam propagates through the atmosphere, it spreads out due to diffraction. It may result in a situation in which the receiver aperture is not able to collect a fraction of the transmitted beam and resulting in beam divergence loss as depicted in Fig. 2.6. A typical FSO system transmits optical beam which is 5–8 cm in diameter at the transmitter. This beam spreads to roughly 1–5 m in diameters after propagating 1 km distance. However, FSO receiver has narrow field of view (FOV), and it is not capable of collecting all the transmitted power resulting in the loss of energy. Figure 2.6 depicts beam divergence loss where the receiver is capable of collecting only a small portion of the transmitted beam.

The optical power collected by the receiver is given by

$$P_R = P_T G_T G_R L_P, \tag{2.7}$$

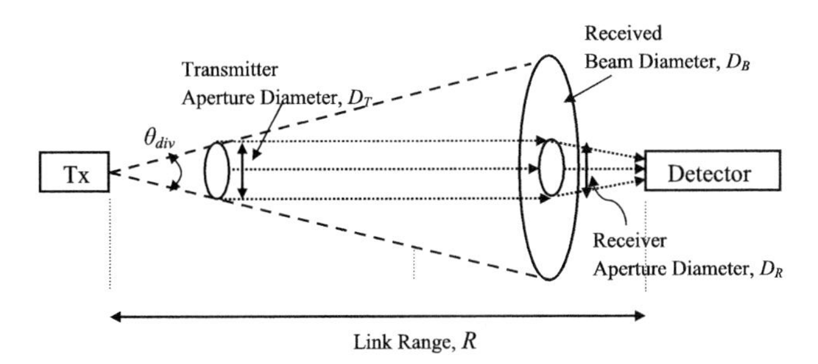

Fig. 2.6 Loss due to beam divergence

where P_T is the transmitted power, L_P the free-space path loss, and G_T and G_R the effective antenna gain of transmitter and receiver, respectively. Substituting the values of $L_P\left[=\left(\frac{\lambda}{4\pi R}\right)^2\right]$, $G_T\left[\approx(4D_T/\lambda)^2\right]$, and $G_R\left[\approx(\pi D_R/\lambda)^2\right]$ (as mentioned in Chap. 1) gives received optical power as

$$P_R \approx P_T\left(\frac{D_T D_R}{\lambda R}\right)^2 \approx P_T\left(\frac{4}{\pi}\right)^2\left(\frac{A_T A_R}{\lambda^2 R^2}\right). \tag{2.8}$$

Therefore, diffraction-limited beam divergence loss/geometric loss expressed in dB is given as

$$L_G(\text{Geometric Loss}) = -10\left[2\log\left(\frac{4}{\pi}\right) + \log\left(\frac{A_T A_R}{\lambda^2 R^2}\right)\right]. \tag{2.9}$$

In general, optical source with narrow beam divergence is preferable. But narrow beam divergence causes the link to fail if there is a slight misalignment between the transceivers. Therefore, an appropriate choice of beam divergence has to be made in order to eliminate the need for active tracking and pointing system and, at the same time, reduce the beam divergence loss. Many times, beam expander is used to reduce the loss due to diffraction-limited beam divergence as beam divergence is inversely proportional to the transmitted aperture diameter ($\theta_{div} \cong \lambda/D_T$). In this case, diffracting aperture is increased with the help of two converging lens as shown in Fig. 2.7.

For non-diffraction-limited case, a source of divergence angle θ_{div} and diameter D_T will make the beam size of $(D_T + \theta_{div}R)$ for link distance equals to R. In this case, the fraction of received power, P_R to the transmitted power, P_T is given by

$$\frac{P_R}{P_T} = \frac{D_R^2}{(D_T + \theta_{div}R)^2}, \tag{2.10}$$

and the beam divergence or geometric loss in dB will be

$$L_G(\text{Geometric Loss}) = -20\log\left[\frac{D_R}{(D_T + \theta_{div}R)}\right]. \tag{2.11}$$

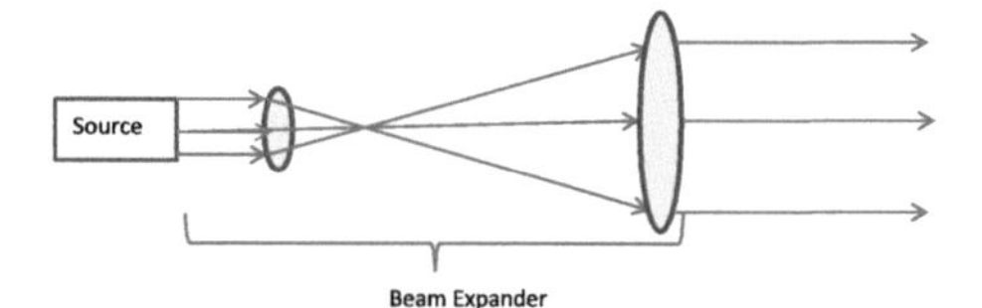

Fig. 2.7 Beam expander to increase diffraction aperture

2.1.1.4 Loss due to Weather Conditions

The performance of FSO link is subject to various environmental factors like fog, snow, rain, etc. that leads to decrease in the received signal power. Out of these environmental factors, the atmospheric attenuation is typically dominated by fog as the particle size of fog is comparable with the wavelength of interest in FSO system. It can change the characteristics of the optical signal or can completely hinder the passage of light because of absorption, scattering, and reflection. The atmospheric visibility is the useful measure for predicting atmospheric environmental conditions. Visibility is defined as the distance that a parallel luminous beam travels through in the atmosphere until its intensity drops 2 % of its original value. In order to predict the optical attenuation statistics from the visibility statistics for estimating the availability of FSO system, the relationship between visibility and attenuation has to be known. Several models that describe the relation between visibility and optical attenuation are given in [9–11]. To characterize the attenuation of optical signal propagating through a medium, a term called "specific attenuation" is used which means attenuation per unit length expressed in dB/km and is given as

$$\beta(\lambda) = \frac{1}{R} \cdot 10\log\left(\frac{P_0}{P_R}\right) = \frac{1}{R} 10\log\left(e^{\gamma(\lambda)R}\right), \quad (2.12)$$

where R is the link length, P_0 the optical power emitted from the transmitter, P_R the optical power at distance R, and $\gamma(R)$ the atmospheric attenuation coefficient. The specific attenuation due to fog, snow, and rain is described below.

(i) Effect of fog: The attenuation due to fog can be predicted by applying Mie scattering theory. However, it involves complex computations and requires detailed information of fog parameters. An alternate approach is based on visibility range information, in which the attenuation due to fog is predicted using common empirical models. The wavelength of 550 nm is usually taken as the visibility range reference wavelength. Equation (2.13) defines the specific attenuation of fog given by common empirical model for Mie scattering.

$$\beta_{fog}(\lambda) = \frac{3.91}{V}\left(\frac{\lambda}{550}\right)^{-p}, \quad (2.13)$$

where V(km) stands for visibility range, λ(nm) is the operating wavelength, and p the size distribution coefficient of scattering.

According to Kim model, p is given as:

$$p = \begin{cases} 1.6 & V > 50 \\ 1.3 & 6 < V < 50 \\ 0.16V + 0.34 & 1 < V < 6 \\ V - 0.5 & 0.5 < V < 1 \\ 0 & V < 0.5 \end{cases} \tag{2.14}$$

According to Kruse model, p is given as:

$$p = \begin{cases} 1.6 & V > 50 \\ 1.3 & 6 < V < 50 \\ 0.585V^{\frac{1}{3}} & V < 6 \end{cases} \tag{2.15}$$

Different weather conditions can be specified based on their visibility range values. Table 2.3 summarizes the visibility range and loss for different weather conditions.

For low visibility weather condition, that is, during heavy fog and cloud, operating wavelength has a negligible effect on the specific attenuation, whereas for light fog and haze when the visibility range is high (6 km), attenuation is quiet less for 1550 nm as compared to 850 and 950 nm. As visibility further increases beyond 20 km (clear weather), dependance of the attenuation on wavelength again decreases. This has been shown in Fig. 2.8a, b obtained by numerical simulation.

(ii) Effect of snow: Attenuation due to snow can vary depending upon the snowflake size and snowfall rate. Since snowflakes are larger in size than raindrop, they produce deeper fades in the signal as compared to the raindrops. Snowflake size can be as large as 20 mm which can completely block the path of the optical signal depending upon the beam width of the signal. For snow,

Table 2.3 Visibility range values corresponding to weather conditions [12]

Weather condition	Visibility range (km)	Loss (dB/km) at 785 nm
Thick fog	0.2	−89.6
Moderate fog	0.5	−34
Light fog	0.770 to 1	−20 to −14
Thin fog/heavy rain (25 mm/hr)	1.9 to 2	−7.1 to −6.7
Haze/medium rain (12.5 mm/hr)	2.8 to 4	−4.6 to −3
Light haze/light rain (2.5 mm/hr)	5.9 to 10	−1.8 to −1.1
Clear/drizzle (0.25 mm/hr)	18 to 20	−0.6 to 0.53
Very clear	23 to 50	−0.46 to −0.21

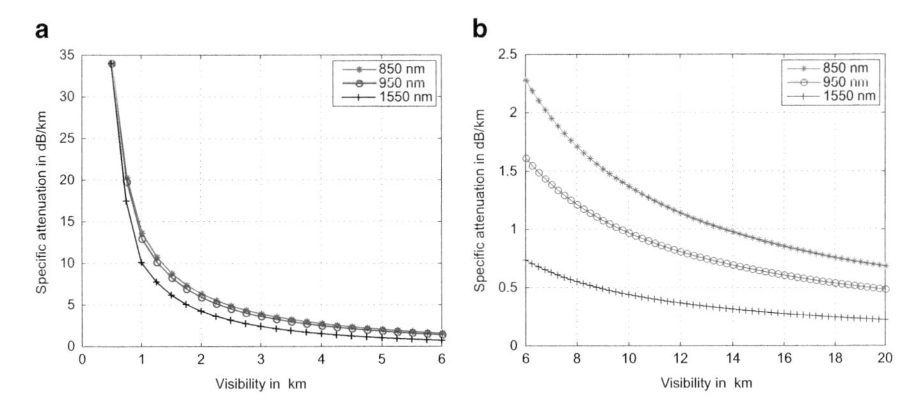

Fig. 2.8 Attenuation vs. visibility. (**a**) For heavy fog and cloud. (**b**) For light fog and haze

attenuation can be classified into dry and wet snow attenuation. The specific attenuation (dB/km) for snow rate S in mm/hr is given by following equation

$$\beta_{snow} = aS^{b}, \tag{2.16}$$

where the values of parameters a and b in dry and wet snow are

$$\begin{aligned}\text{Dry snow}: \ a &= 5.42 \times 10^{-5} + 5.4958776, \ \ b = 1.38 \\ \text{Wet snow}: \ a &= 1.023 \times 10^{-4} + 3.7855466, \ b = 0.72\ .\end{aligned} \tag{2.17}$$

The snow attenuation based on visibility range can be approximated by the following empirical model

$$\alpha_{snow} = \frac{58}{V} \tag{2.18}$$

(iii) Effect of rain: The sizable rain droplets can cause wavelength-independent scattering, and the attenuation produced by rainfall increases linearly with rainfall rate. The specific attenuation for rain rate **R** (mm/hr) is given by

$$\beta_{rain} = 1.076\mathbf{R}^{0.67} \tag{2.19}$$

The rain attenuation for FSO links can be reasonably well approximated by empirical formula and is given by

$$\alpha_{rain} = \frac{2.8}{V} \tag{2.20}$$

where V is visibility range in km and its values based on rainfall rate is summarized in Table 2.4.

Table 2.4 Rainfall rates and their visibility ranges [13]

Rainfall type	Rainfall rate, R (mm/hr)	Visibility range, V (km)
Heavy rain	25	1.9–2
Medium rain	12.5	2.8–40
Light rain/drizzle	0.25	18–20

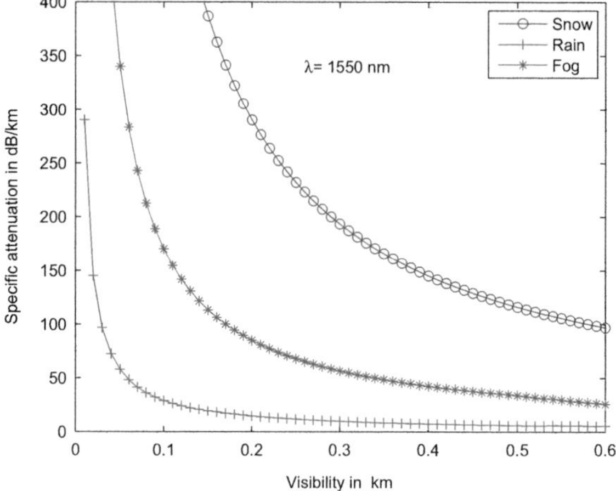

Fig. 2.9 Attenuation for fog, snow and rain

It is clear from Fig. 2.9 that the attenuation is maximum for snow and lesser for rain as compared to fog. In case of fog, there is a sudden increase in attenuation for visibility range less than 150 m. Visibility range <150 m corresponds to heavy fog and cloudy weather.

2.1.1.5 Pointing Loss

The loss that occurs due to imperfect alignment between transmitter and receiver is called pointing loss. A large pointing loss can lead to intolerable signal fades and can significantly degrade the system performance. It is due to the fact that the random platform jitter is generally much larger than the transmitter beam width. Hence, a very tight acquisition, tracking, and pointing (ATP) subsystem is required to reduce the loss due to misalignment. The transmitted power that is allocated for the pointing of the optical beam is not used for communication. Therefore, it is highly desirable to keep the pointing loss due to misalignment of the narrow laser beam as small as possible so that the sufficient power is available for communication. It is desirable to achieve sub-microradian pointing system by using inertial sensors, focal plane arrays, and the steering mirror. The pointing loss of the transmitter that must be considered for the link budget analysis is given as

$$L_p = \exp\left(\frac{-8\theta_{jitter}^2}{\theta_{div}^2}\right), \tag{2.21}$$

where θ_{jitter} is the beam jitter angle and θ_{div} the transmitter beam divergence as shown in Fig. 2.6.

2.1.2 *Atmospheric Turbulence*

The atmosphere can be thought as a viscous fluid that has two distinct states of motion, i.e., laminar and turbulent. In laminar flow, the velocity flow characteristics are uniform, or they change in some regular fashion. In case of turbulent flow, the velocity loses its uniform characteristics due to dynamic mixing and acquire random sub-flows called turbulent eddies. The transition from laminar to turbulent motion is determined by the nondimensional quantity called Reynolds number. This number is defined as $\mathrm{Re} = \mathbf{V}l_f/\nu_k$, where $\mathbf{V}$ is the characteristic velocity (in m/s), l_f the dimension of flow (in m), and ν_k the kinematic viscosity (in m^2/s). If the Reynolds number exceeds the "critical Reynolds Number," the flow is considered turbulent in nature. Near the ground, the characteristic scale size l_f is approximately 2 m, wind velocity is 1 to 5 m/s, and ν_k is about $0.15 \times 10^{-4}\,\mathrm{m}^2/\mathrm{s}$. This gives a large value for the Reynolds number of the order of $\mathrm{Re} \sim 10^5$. Therefore, close to the ground level, the flow may be considered to be highly turbulent.

In order to understand the structure of atmospheric turbulence, it is convenient to adopt the energy cascade theory of turbulence [14, 15]. As per this theory, when the wind velocity is increased, the Reynolds number exceeds the critical value. This results in the local unstable air masses called turbulent eddies with their characteristic dimensions slightly smaller than, and independent of, the parent flow [16]. Under the influence of the inertial forces, the larger size eddies break up into smaller eddies until the inner scale size l_0 is reached. The family of eddies bounded above by the outer scale L_0 and below by the inner scale l_0 forms the inertial subrange. The outer scale denotes the scale below which the turbulent properties are independent of the parent flow. Typically, the outer scale is in the order of about 10 to 100 m and is usually assumed to grow linearly with the height above the ground. The inner scale is of the order of 1 to 10 mm near the ground, but it is of the order of centimeters or more in the troposphere and stratosphere. Scale size smaller than the inner scale l_0 belongs to the viscous dissipation range. In this range, the turbulent eddies disappear, and the remaining energy is dissipated in the form of heat. This phenomenon is known as Kolmogorov theory [17] of turbulence and is depicted in Fig. 2.10.

The mathematical model of the atmospheric turbulence and its effects on the optical beam propagation assumes that the fluctuations in the atmospheric parameter are stationary random processes having statistically homogeneous and isotropic nature. Within this mathematical framework, the structure function in the inertial range satisfies the universal two-thirds power law, i.e., it follows $r^{2/3}$ dependence, where r refers to the spatial scale defined as

$$r = \left|\vec{r_1} - \vec{r_2}\right|, \qquad l_0 \leq r \leq L_0, \tag{2.22}$$

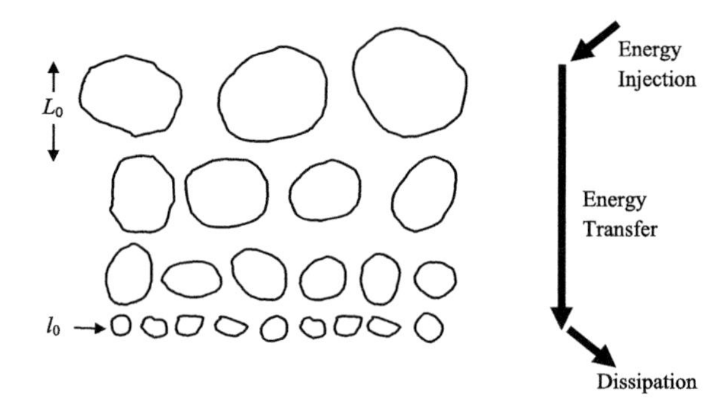

Fig. 2.10 Kolmogorov model where L_0 and l_0 are the outer and inner scale of turbulent eddies, respectively

where $\overrightarrow{r_1}$ and $\overrightarrow{r_2}$ refer to position vectors at two points separated by distance r in space. The structure function for a random variable $x(r)$ is then given as

$$\begin{aligned} D_x\left(r_{(\cdot)}\right) &= D_x\left(f\left(\overrightarrow{r_1}, \overrightarrow{r_2}\right)\right) \\ &= \left\langle \left| x\left(\overrightarrow{r_1}\right) - x\left(\overrightarrow{r_2}\right)\right|^2 \right\rangle. \end{aligned} \tag{2.23}$$

The random variable $x(r)$ is assumed to have a mean and a superimposed fluctuating component that is represented as

$$x(r) = \left\langle x\left(\overrightarrow{r}\right)\right\rangle + x'\left(\overrightarrow{r}\right). \tag{2.24}$$

In the above equation, the first term in the angle bracket is the slowly varying mean component and second term the random fluctuations. Using Eq. (2.24), the structure function in Eq. (2.23) can be rewritten as

$$D_x\left(r_{(\cdot)}\right) = \left[\left\langle x\left(\overrightarrow{r_1}\right)\right\rangle - \left\langle x\left(\overrightarrow{r_2}\right)\right\rangle\right]^2 + \left\langle \left[\left\langle x'\left(\overrightarrow{r_1}\right)\right\rangle - \left\langle x'\left(\overrightarrow{r_2}\right)\right\rangle\right]^2 \right\rangle. \tag{2.25}$$

The first term in Eq. (2.25) becomes zero for a stationary random process. This makes structure function a useful parameter for describing the random fluctuations. With the atmosphere as the propagation medium, these random fluctuations can be associated with any of these parameters, i.e., velocity, temperature, and refractive index. The structure function for wind velocity $D_v\left(r_v\right)$ is given as [16]

$$D_v\left(r_v\right) = \left\langle \left(v_1 - v_2\right)^2 \right\rangle = C_v^2 r^{2/3}, \qquad l_0 \le r \le L_0, \tag{2.26}$$

where v_1 and v_2 are the velocity components at two points separated by distance r and C_v^2 the velocity structure constant (in $\mathrm{m}^{4/3}/\mathrm{s}^2$) that measures the amount of energy in the turbulence. Similarly, the structure function for the temperature is given as $D_t\left(r_t\right) = C_t^2 r^{2/3}$, where C_t^2 is the temperature structure constant (in $\mathrm{deg}^2/\mathrm{m}^{2/3}$). The turbulence in the atmosphere also results from the random fluctuations of the atmospheric refractive index n due to variations in temperature and pressure along the propagation path in the atmosphere. In general, the refractive index of the atmosphere at any point r in space can be expressed as sum of the average and the fluctuating terms, i.e.,

$$n(r) = n_0 + n'(r), \tag{2.27}$$

where $n_0 = \langle n(r) \rangle \approx 1$ is the mean value of index of refraction and $n'(r)$ represents the random deviation of $n(r)$ from its mean value. Therefore, the above equation can be rewritten as

$$n(r) = 1 + n'(r). \tag{2.28}$$

The index of refraction of the atmosphere is related to temperature and pressure of the atmosphere and is given as

$$\begin{aligned} n(r) &= 1 + 7.66 \times 10^{-6} \left(1 + 7.52 \times 10^{-3}\lambda^{-2}\right) \frac{P'(r)}{T'(r)} \\ &\cong 1 + 79 \times 10^{-6} \left(\frac{P'(r)}{T'(r)}\right), \end{aligned} \tag{2.29}$$

where λ is the wavelength in µm, P' the atmospheric pressure in mbar, and T' the temperature of the atmosphere in Kelvin. Changes in the optical signal due to absorption or scattering by the molecules or aerosols are not considered here. The structure function for refractive index, $D_n(r)$, can be expressed as $D_n(r_n) = C_n^2 r^{2/3}$, where C_n^2 is called refractive index structure constant [16] and is a measure of the strength of fluctuations in the refractive index. C_n^2 is related to temperature structure constant, C_t^2 as

$$C_n^2 = \left[79 \times 10^{-6} \frac{P'}{T'^2}\right]^2 C_t^2. \tag{2.30}$$

The C_t^2 is determined by taking the measurements of mean square temperature between two points separated by a certain distance along the propagation path (in $\text{deg}^2/\text{m}^{2/3}$). The other parameters are as defined earlier. It is obvious from Eq. (2.30) that refractive index structure parameter can be obtained by measuring temperature, pressure, and temperature spatial fluctuations along the propagation path. All the expressions for the structure function are defined for the inertial subrange, i.e., for $l_0 \ll r \ll L_0$. Out of all the structure constants referred above, the refractive index structure constant, i.e., C_n^2, is considered the most critical parameter along the propagation path in characterizing the effects of atmospheric turbulence.

Depending on the size of turbulent eddy and transmitter beam size, three types of atmospheric turbulence effects are observed:

- Beam Wander (or beam steering): If the size of eddies are larger than the transmitter beam size, it will deflect the beam as a whole in random manner from its original path. This phenomenon is called beam wander, and it effectively leads to pointing error displacement of the beam that causes the beam to miss the receiver.
- Beam Scintillation: If the eddy size is of the order of beam size, then the eddies will act like lens that will focus and de-focus the incoming beam leading to irradiance fluctuations at the receiver and the process is called scintillation. Scintillations causesthe loss of signal-to-noise ratio and result in deep random signal fades. The effect of scintillation can be reduced by employing techniques like multiple transmit/receive antennae, aperture averaging, etc.
- Beam Spreading: If the eddy size is smaller than the beam size, then a small portion of the beam will be diffracted and scattered independently. This will lead to reduction in the received power density and will also distort the received wavefront. However, the effect of turbulence-induced beam spreading will be negligible if the transmitter beam diameter is kept smaller than the coherence length of the atmosphere [18] or if the receiver aperture diameter is kept greater than the size of first Fresnel zone $\sqrt{R/k}$ [19]. In this case, the only effect will be due to turbulence-induced beam wander effect and scintillation effect.

2.1.2.1 The Effect of Beam Wander

Beam wander is the random deflection of the optical beam as it propagates through the large-scale inhomogeneities present in the turbulent atmosphere. In case of FSO uplink communication from ground to satellite, when the target is in the far field and turbulence exists in the near field of the transmitter, beam diameter is often smaller than the outer scale of the turbulence. In this case, the Gaussian profile of an optical beam gets highly skewed over short duration after propagating through turbulent atmosphere. In this process, the instantaneous point of maximum irradiance, known as "hot spot," gets displaced from its on-axis position. The movement of hot spot and short-term beam centroid will effectively lead to outer large circle over a long period of time and is called as long-term spot size. The long-term spot size is the result of beam wander, free-space diffraction spreading, and additional spreading due to small-scale turbulent eddies smaller than beam size. The long-term spot size is therefore given as [20–22]

$$W_{LT}^2(R) = \underbrace{W^2(R) + W^2(R)\,T_{ss}}_{\text{Beam Spread, } W_{ST}^2} + \underbrace{\langle r_c^2 \rangle}_{\text{Beam Wander}}, \tag{2.31}$$

where $W(R)$ is the beam size after propagating distance R and T_{ss} describes the short-term spread of the beam due to atmospheric turbulence. The combined movement of the hot spot and short-term beam as depicted in Fig. 2.11a leads to the large outer circle over a long time period, and it is called long-term spot size W_{LT} as

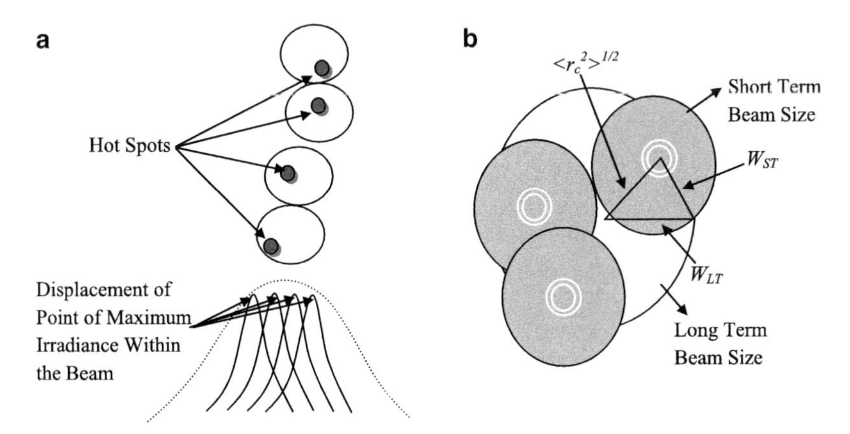

Fig. 2.11 Beam wander effect described by (**a**) Movement of the "hot spot" within the beam and (**b**) Beam wander variance $\langle r_c^2 \rangle^{1/2} = \sqrt{W_{LT}^2 - W_{ST}^2}$, where W_{ST} is the short-term beam radius and W_{LT} the long-term beam radius at the receiver (the *shaded circles* depict random motion of the short-term beam in the receiver plane [16])

shown in Fig. 2.11b. Therefore, the resultant long-term spot size is the superposition of the instantaneous spots that reach the receiver. The first term in Eq. (2.31) is due to free-space diffraction, middle term the additional spreading by the turbulent eddies of size smaller than the beam size, and last term the beam wander displacement variance caused by the large-size turbulent eddies. By making use of appropriate filters that only permit random inhomogeneities of size equal to or greater than beam size, the effect due to small-scale spread will be eliminated, and only contribution will be due to beam wander effect. *It should be noted that beam wander effect is negligible in case of downlink signals from the satellites.* This is because the beam size when reaches the atmosphere is much larger than the turbulent eddy size and that would not displace the beam centroid significantly. Instead, the wavefront tilt at the receiver produced by the atmospheric turbulence gives rise to angle of arrival fluctuations. The beam wander displacement variance $\langle r_c^2 \rangle$ for a collimated uplink beam is given as [22]

$$\langle r_c^2 \rangle = 7.25\,(H - h_0)^2 \sec^3(\theta)\, W_0^{-1/3} \int_{h_0}^{H} C_n^2(h) \left(1 - \frac{h - h_0}{H - h_0}\right)^2 dh, \tag{2.32}$$

where $C_n^2(h)$ is the refractive index structure parameter, θ the zenith angle, W_0 the transmitter beam size, H and h_0 are the altitude of satellite and transmitter, respectively. For ground-based transmitter, $h_0 = 0$, and satellite altitude $H = h_0 + R\cos(\theta)$. Equation (2.32) can now be rewritten as [22]

$$\langle r_c^2 \rangle \approx 0.54\,(H - h_0)^2 \sec^2(\theta) \left(\frac{\lambda}{2W_0}\right)^2 \left(\frac{2W_0}{r_0}\right)^{5/3}, \tag{2.33}$$

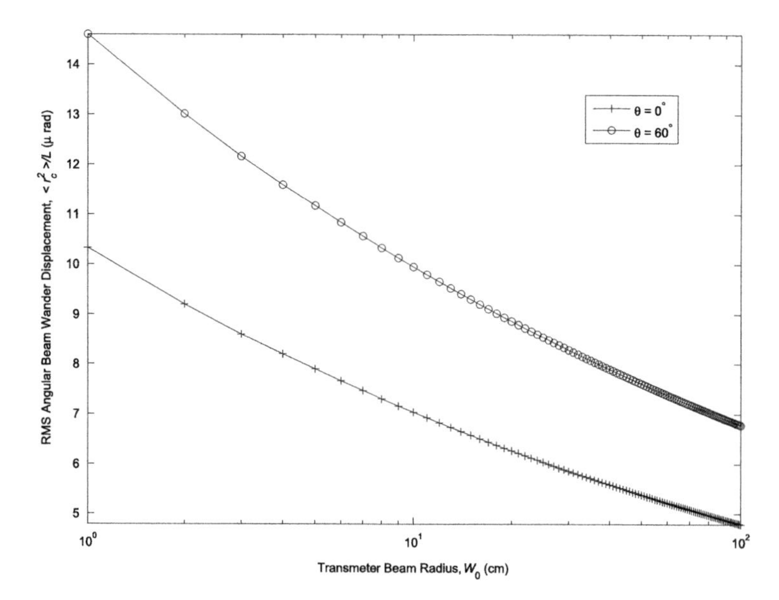

Fig. 2.12 The rms angular beam wander variance as a function of transmitter beam radius for ground-to-satellite FSO link

where r_0 is the atmospheric coherence length (i.e., Fried parameter) and is given as

$$r_0 = \begin{cases} \left(1.46 C_n^2 k^2 R\right)^{-3/5} & \text{for horizontal link} \\ \left[0.423 k^2 \sec(\theta) \int_{h_0}^{H} C_n^2(h)\, dh\right]^{-3/5} & \text{for vertical link.} \end{cases} \tag{2.34}$$

It is seen from Eqs. (2.33) and (2.34) that beam wander variance involves the free-space diffraction angle ($\lambda/2W_0$) and tilt phase fluctuations averaged over the transmitter aperture and is of the order of $(2W_0/r_0)^{5/3}$. Variations of $\langle r_c^2 \rangle$ with W_0 for zenith angles ($\theta = 0°$ and $60°$) and $R = 40{,}000$ km are shown in Fig. 2.12. It is seen from this figure that the angular beam wander displacement variance is large for small beam size W_0 and decreases rapidly with the increase in the value of W_0.

The beam wander effect causes widening of the long-term beam profile near the boresight that flattens the top of the beam as shown in Fig. 2.13. The flattened beam profile effectively leads to effective pointing error of the beam σ_{pe}, and this results in the increase of scintillation index σ_I^2, discussed in the next section. In other words, the turbulence-induced beam wander effect effectively leads to pointing error displacement that can significantly affect the channel.

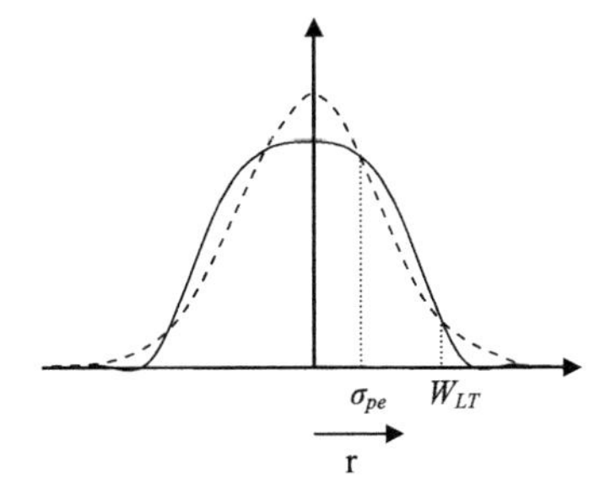

Fig. 2.13 Flattened beam profile as a function of radial displacement that leads to effective pointing error σ_{pe}

2.1.2.2 The Scintillation Effect

The irradiance fluctuations within the cross section of the received beam after propagating through turbulent atmosphere are commonly referred as "scintillation" and are measured in terms of scintillation index (or normalized variance of irradiance) σ_I^2. It causes loss of signal-to-noise ratio and induces deep signal fades. The σ_I^2 is defined as [23]

$$\sigma_I^2 = \frac{\langle I^2 \rangle - \langle I \rangle^2}{\langle I \rangle^2} = \frac{\langle I^2 \rangle}{\langle I \rangle^2} - 1, \tag{2.35}$$

where I is the irradiance (intensity) at some point in the detector plane and the angle bracket $\langle\rangle$ denotes an ensemble average. For X amplitude of the transmitted optical beam, the received irradiance, I, at the receiver takes the form as

$$I = I_0 \exp\left[2X - 2E\left[X\right]\right], \tag{2.36}$$

where I_0 is the intensity without turbulence. From Eqs. (2.35) and (2.36), σ_I^2 in terms of log-amplitude variance, σ_x^2 is given by

$$\sigma_I^2 \approx 4\sigma_x^2, \ \text{for}\, \sigma_x^2 << 1. \tag{2.37}$$

Further, the variance of log irradiance (also called Rytov variance) σ_R^2 is related to σ_I^2 as below

$$\sigma_I^2 = \exp\left(\sigma_R^2\right) - 1 \approx \sigma_R^2 \,\text{for}\, \sigma_R^2 << 1. \tag{2.38}$$

In weak turbulence, scintillation index is expressed as

$$\sigma_I^2 = \sigma_R^2 = 1.23 C_n^2 k^{7/6} R^{11/6} \,\text{for plane waves} \tag{2.39}$$

and

$$\sigma_I^2 = 0.4\sigma_R^2 = 0.5 C_n^2 k^{7/6} R^{11/6} \,\text{for spherical waves}, \ \text{respectively}, \tag{2.40}$$

where k is wave number $(=2\pi/\lambda)$. It is clear from Eqs. (2.39) and (2.40) that for a weak turbulence conditions, longer wavelength will experience lesser irradiance fluctuations for a given link range. Scintillation index for strong turbulence is given by

$$\sigma_I^2 = \begin{cases} 1 + \frac{0.86}{\sigma_R^{4/5}}, \ \sigma_R^2 >> 1 & \text{for plane wave} \\ \\ 1 + \frac{2.73}{\sigma_R^{4/5}}, \ \sigma_R^2 >> 1 & \text{for spherical wave} \end{cases} \tag{2.41}$$

Clearly, Eq. (2.41) shows that for strong turbulence, smaller wavelength will experience lesser irradiance fluctuations.

Various studies have been performed to develop the mathematical model for the probability density function (PDF) of the randomly fading received signal irradiance. These studies have led to various statistical models that can describe the turbulence-induced scintillation over a wide range of atmospheric conditions.

For weak turbulence, $\sigma_I^2 < 1$, and the irradiance statistics is given by lognormal model. This model is widely used due to its simplicity in terms of mathematical computations. The PDF of the received irradiance, I, is given as

$$f(I) = \frac{1}{\sqrt{2\pi\sigma_I^2 I}} \exp\left[-\frac{(ln(I) - \mu)^2}{2\sigma_I^2}\right], \tag{2.42}$$

where μ is the mean of ln (I). Since $\sigma_I^2 = 4\sigma_x^2$, the above lognormal pdf can be rewritten as

$$f(I) = \frac{1}{2\sqrt{2\pi\sigma_x^2 I}} \exp\left[-\frac{(ln(I) - \mu)^2}{8\sigma_x^2}\right]. \tag{2.43}$$

When the strength of turbulence increases, lognormal pdf shows large deviation as compared to experimental data. Therefore, lognormal statistics is not appropriate model in case of strong fluctuation regimes.

For strong turbulence, $\sigma_I^2 \geq 1$, and the field amplitude is Rayleigh distributed which in turn leads to negative exponential statistics for received irradiance [2]. Its pdf is given by

$$f\,(I) = \frac{1}{I_0} \exp\left(-\frac{I}{I_0}\right), \ I \geq 0, \tag{2.44}$$

where I_0 is the mean irradiance. In this case, $\sigma_I^2 \approx 1$, and this happens only far into saturation regime.

Besides these two models, a number of other statistical models [24] are there in the literature to describe the scintillation statistics in either a regime of strong turbulence (K model) or all the regimes (I-K and gamma-gamma [25] models). For $3 < \sigma_I^2 < 4$, the intensity statistics is given by K distribution. Thus K distribution describes only the strong turbulence intensity statistics. This model was originally proposed for non-Rayleigh sea echo, but later it was discovered that it is an appropriate model for characterizing the amplitude fluctuations in strong atmospheric conditions. Its pdf is given by

$$f(I) = \frac{2}{\Gamma(\alpha)} \alpha^{\frac{\alpha+1}{2}} I^{\frac{\alpha-1}{2}} K_{\alpha-1}\left(2\sqrt{\alpha I}\right), \quad I > 0, \tag{2.45}$$

where α is a channel parameter related to the effective number of discrete scatterers and $\Gamma(\cdot)$ is the well-known gamma function. When $\alpha \rightarrow \infty$, the gamma distribution function approaches delta function, and K distribution reduces to negative exponential distribution. However, K distribution lacked the numerical computation in closed form. Also, it cannot easily relate the mathematical parameters with the observables of atmospheric turbulence, and therefore, it limits the applicability and utilization.

Another generalized form of K distribution that is applicable to all conditions of atmospheric turbulence, including weak turbulence for which the K distribution is not theoretically applicable, is $I - K$ distribution. In this case, the field of optical wave is modeled as sum of coherent (deterministic) component and a random component. The intensity is assumed to be governed by the generalized Nakagami distribution. The pdf of $I - K$ distribution is given by [26]

$$f(I) = \begin{cases} 2\alpha(1+\rho)\left(1+\frac{1}{\rho}\right)^{\frac{\alpha-1}{2}} I^{\frac{\alpha-1}{2}} K_{\alpha-1}\left(2\sqrt{\alpha\rho}\right) I_{\alpha-1}\left(2\sqrt{\alpha(1+\rho)I}\right), & 0 < I < \frac{\rho}{1+\rho} \\ 2\alpha(1+\rho)\left(1+\frac{1}{\rho}\right)^{\frac{\alpha-1}{2}} I^{\frac{\alpha-1}{2}} I_{\alpha-1}\left(2\sqrt{\alpha\rho}\right) K_{\alpha-1}\left(2\sqrt{\alpha(1+\rho)I}\right) & I > \frac{\rho}{1+\rho} \end{cases}, \tag{2.46}$$

where $I_a(\cdot)$ is the modified Bessel function of first kind of order a. The normalized $I - K$ distribution in the above equation involves two empirical parameters ρ and α whose values are selected by matching the first three normalized moments of the distribution. The parameter ρ is a measure of the power ratio of mean intensities of the coherent and random components of the field.The value of ρ is relatively large for extremely weak turbulence. By properly selecting the values of α and ρ, both weak and strong turbulence can be obtained. Because of the symmetry of the functional forms involving the I and K Bessel functions, this distribution is henceforth referred to as the $I - K$ distribution. The $I - K$ distribution reduces to the K distribution in the limit $\rho \rightarrow 0$.

However, $I - K$ distribution is difficult to express in closed-form expressions. In that case, the gamma-gamma distribution is used to successfully describe the scintillation statistics for weak to strong turbulence. In this model, the normalized irradiance, I, is defined as the product of two independent random variables, i.e., $I = I_X I_Y$, where I_X and I_Y represent large-scale and small-scale turbulent eddies and each of them following a gamma distribution. This leads to gamma-gamma distribution whose pdf is given as

$$f_I(I) = \frac{2(\alpha\beta)^{(\alpha+\beta)/2}}{\Gamma(\alpha)\Gamma(\beta)} I^{((\alpha+\beta)/2)-1} K_{\alpha-\beta}\left(2\sqrt{\alpha\beta I}\right), \quad I > 0, \tag{2.47}$$

where $K_a(\cdot)$ is the modified Bessel function of second kind of order a. The parameters α and β are the effective number of small-scale and large-scale eddies of the scattering environment and are related to the atmospheric conditions through the following expressions

$$\alpha = \left[\exp\left[\frac{0.49\chi^2}{\left(1 + 0.18d^2 + 0.5\chi^{12/5}\right)^{7/6}}\right] - 1\right]^{-1} \tag{2.48}$$

and

$$\beta = \left[\exp\left[\frac{0.51\chi^2\left(1 + 0.69\chi^{12/5}\right)^{-5/6}}{\left(1 + 0.9d^2 + 0.62d^2\chi^{12/5}\right)^{7/6}}\right] - 1\right],^{-1} \tag{2.49}$$

where $\chi^2 = 0.5C_n^2 k^{7/6} R^{11/6}$ and $d = \left(kD_R^2/4R\right)^{1/2}$. The parameter $k = \frac{2\pi}{\lambda}$ is the optical wave number, D_R the diameter of the receiver collecting lens aperture, and R the link range in meters, C_n^2 refractive index structure parameter whose value varies from 10^{-13} m$^{-2/3}$ for strong turbulence to 10^{-17} m$^{-2/3}$ for weak turbulence. Since the mean value of this turbulence model is $E[I] = 1$ and the second moment is given by $E\left[I^2\right] = (1 + 1/\alpha)(1 + 1/\beta)$, therefore, scintillation index (SI), that gives the strength of atmospheric fading, is defined as

$$SI = \frac{E\left[I^2\right]}{(E[I])^2} - 1 = \frac{1}{\alpha} + \frac{1}{\beta} + \frac{1}{\alpha\beta}. \tag{2.50}$$

While still mathematically complex, the gamma-gamma distribution can be expressed in closed form and can relate to various values of scintillation index unlike *I-K* distribution [2]. Figure 2.14 shows the range of scintillation index for various types of distribution used to model the intensity statistics.

Another turbulence model proposed in [27] is double generalized gamma (double GG) distribution which is suitable for all regimes of turbulence, and it covers almost all the existing statistical models of irradiance fluctuations as special cases.

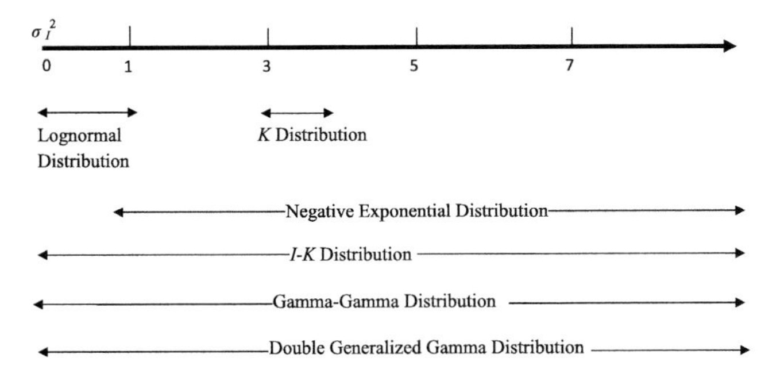

Fig. 2.14 Various distributions for intensity statistics

2.1.3 Effect of Atmospheric Turbulence on Gaussian Beam

Let us consider a Gaussian beam with amplitude A_0 propagating in free space and the transmitter located at $z = 0$. The amplitude distribution in this plane is a Gaussian function with effective beam radius W_0 defined as the radius at which the field intensity falls off to $1/e$ of that on the beam axis. The Gaussian beam at $z = 0$ is described by

$$U_0(r, 0) = A_0 \exp\left(-\frac{r^2}{W_0^2} - i\frac{kr^2}{2F_0}\right), \tag{2.51}$$

where r is the distance from beam center line in the transverse direction, k $(= 2\pi/\lambda)$ the optical wave number as defined earlier, and F_0 the phase front radius of curvature which specifies the beam forming. The cases $F_0 > 0$, $F_0 = \infty$, and $F_0 < 0$ correspond to converging, collimated, and diverging beam forms, respectively, [28] as shown in Fig. 2.15.

For a propagation path of range R along positive z axis, the free-space Gaussian beam wave is described as [22]

$$U_0(r, R) = \frac{A_0}{\Theta_0 + i\Lambda_0} \exp\left(ikR - \frac{r^2}{W^2} - i\frac{kr^2}{2F'}\right). \tag{2.52}$$

The parameters Θ_0 and Λ_0 are referred to as transmitter beam parameters as they are defined in terms of beam characteristics at the transmitter and are given as

$$\Theta_0 = 1 - \frac{R}{F_0}, \quad \Lambda_0 = \frac{2R}{kW_0^2}. \tag{2.53}$$

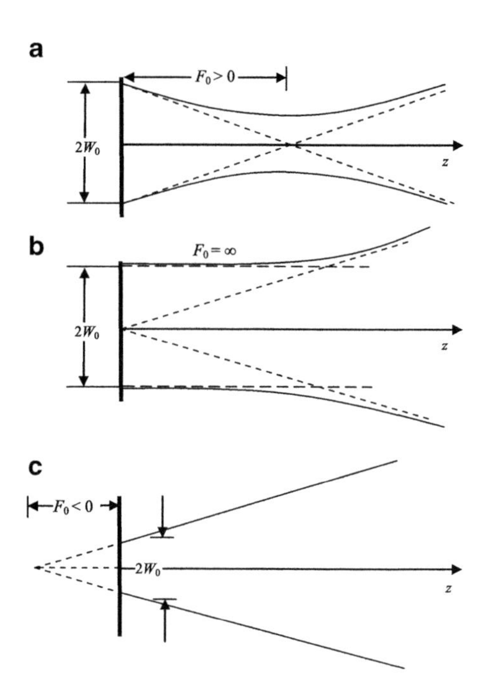

Fig. 2.15 Representation of (**a**) convergent beam, (**b**) collimated beam, and (**c**) divergent beam, respectively

The parameter Θ_0 represents the amplitude change in the wave due to focusing (refraction) and Λ_0 the amplitude change due to diffraction. The parameter Θ_0 is also called curvature parameter and Λ_0 the Fresnel ratio at the transmitter plane. The other parameters W and in Eq. (2.52) are the effective beam radius and phase front radius of curvature of the beam, respectively, at the receiver plane. With transmitter beam parameters as described above, the receiver beam parameters are given by

$$\Theta = 1 + \frac{R}{F'} = \frac{\Theta_0}{\Theta_0^2 + \Lambda_0^2}, \tag{2.54}$$

$$\Lambda = \frac{2R}{kW^2} = \frac{\Lambda_0}{\Theta_0^2 + \Lambda_0^2}. \tag{2.55}$$

Like their counterparts, Θ_0 and Λ_0, the receiver parameter Θ describes the focusing or refractive effect on the amplitude of the wave and Λ the diffraction effects on the amplitude (i.e., diffractive spreading of the wave). The parameters W and F' are related to beam parameters as [22]

$$W = W_0 \left(\Theta_0 + \Lambda_0\right)^{1/2} = \frac{W_0}{\left(\Theta^2 + \Lambda^2\right)^{1/2}}, \tag{2.56}$$

$$F' = \frac{F_0\left(\Theta^2 + \Lambda^2 - \Theta\right)}{\left(\Theta - 1\right)\left(\Theta^2 + \Lambda^2\right)} = \frac{F_0\left(\Theta_0^2 + \Lambda_0^2\right)\left(\Theta_0 - 1\right)}{\Theta_0^2 + \Lambda_0^2 - \Theta_0}. \tag{2.57}$$

The free-space irradiance profile (without atmospheric turbulence) at the receiver plane is the square magnitude of the field given in Eq. (2.52) and is given as

$$I_0\,(r, R) = \mid U_0\,(r, R)\mid^2. \tag{2.58}$$

From Eqs. (2.52) and (2.58), we get

$$I_0\,(r, R) = \left(\frac{A_0^2}{\Theta_0^2 + \Lambda_0^2}\right)\exp\left(-\frac{2r^2}{W^2}\right) \quad (\mathrm{W/m^2}). \tag{2.59}$$

Since $\sqrt{\Theta^2 + \Lambda^2} = 1/\sqrt{\Theta_0^2 + \Lambda_0^2}$, the above equation can be written as

$$I_0\,(r, R) = A_0^2\left(\Theta^2 + \Lambda^2\right)\exp\left(-\frac{2r^2}{W^2}\right). \tag{2.60}$$

The on-axis irradiance is $I_0\,(0, R) = \left(\frac{A_0^2}{\Theta_0^2+\Lambda_0^2}\right) = A_0^2\left(\Theta^2 + \Lambda^2\right)$. Figure 2.16 shows the uplink Gaussian beam profile illustrating the spot size $W(R)$, mean transmit intensity $I_0\,(r, 0)$, mean received intensity $I_0\,(r, R)$, and angular pointing error α_r.

When a Gaussian beam propagates through the turbulent atmosphere, it experiences various deleterious effects including beam spreading, beam wander, and beam scintillation that will result in variation of Gaussian beam parameters. For weak atmospheric fluctuations, the solution of wave equation for various propagation problems can be analyzed by conventional Rytov approximation. However, this

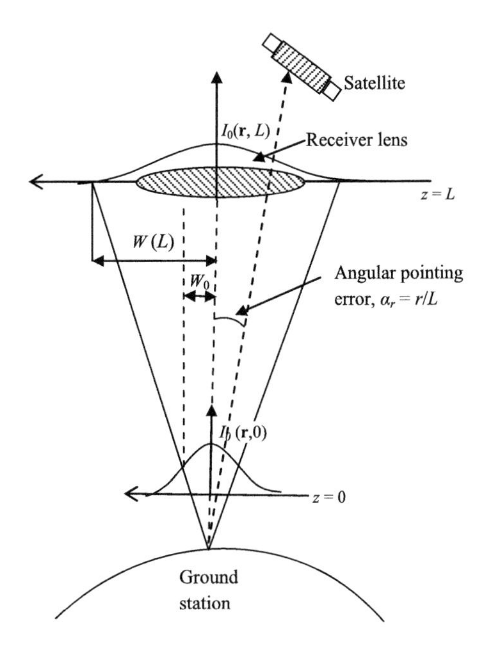

Fig. 2.16 Gaussian beam profile parameters for uplink propagation path

approximation is limited to weak fluctuation conditions as it does not account for decreasing transverse spatial coherence radius of the propagating wave. Therefore, a relatively simple model for irradiance fluctuations for Gaussian beam wave that is applicable in weak to strong turbulence has been suggested by modifying the conventional Rytov approximation. The modified Rytov approximation can be used to describe the irradiance fluctuations throughout the propagation path near the ground and allows us to extend the weak fluctuation results to moderate/strong fluctuation regime. In the following subsections, the Gaussian beam propagation using conventional Rytov approximation is discussed, and then the discussion is extended to modified Rytov approximation.

2.1.3.1 Conventional Rytov Approximation

In the conventional Rytov approximation, it is assumed that the optical field at link range R from the transmitter is given by

$$\begin{aligned} U(r,R) &= U_0(r,R)\exp\left[\Psi(r,R)\right] \\ &= U_0(r,R)\exp\left[\Psi_1(r,R)+\Psi_2(r,R)+\ldots\right], \end{aligned} \tag{2.61}$$

where $U_0(r,R)$ is the free-space diffraction-limited Gaussian beam wave at the receiver given by Eq. (2.52) and $\Psi(r,R)$ the complex phase fluctuations of the field due to random refractive index inhomogeneities along the propagation path R. The parameters $\Psi_1(r,R)$ and $\Psi_2(r,R)$ are the first-order and second-order perturbations, respectively. The statistical moments of the optical field given in Eq. (2.61) involve the ensemble average of the first- and second-order perturbations. The first-order moment is given as

$$\langle U(r,R)\rangle = U_0(r,R)\langle\exp\left[\Psi(r,R)\right]\rangle. \tag{2.62}$$

The second-order moment, also called mutual coherence function [16] (MCF), is given by

$$\begin{aligned} \Gamma_2(r_1,r_2,R) &= \langle U(r_1,R)\,U^*(r_2,R)\rangle \\ &= U_0(r_1,R)\,U_0^*(r_2,R)\left\langle\exp\left[\Psi(r_1,R)+\Psi^*(r_2,R)\right]\right\rangle, \end{aligned} \tag{2.63}$$

where * denotes complex conjugate and notation $\langle\rangle$denotes the ensemble average which can be calculated using the following equation:

$$\langle\exp(\Psi)\rangle = \exp\left[\langle\Psi\rangle+\frac{1}{2}\left(\left\langle\Psi^2\right\rangle-\langle\Psi\rangle^2\right)\right]. \tag{2.64}$$

These ensemble averages can be expressed as linear combinations of integrals designated by $E_1(0,0)$, $E_2(r_1,r_2)$, and $E_3(r_1,r_2)$ [16]. In particular,

$$\begin{aligned}\langle\exp\left[\Psi\left(r,R\right)\right]\rangle &= \langle\,\exp\left[\Psi_1\left(r,R\right)+\Psi_2\left(r,R\right)\right]\rangle\\ &= \exp\left[E_1\left(0,0\right)\right],\end{aligned} \tag{2.65}$$

$$\begin{aligned}\left\langle\exp\left[\Psi\left(r_1,R\right)+\Psi^*\left(r_2,R\right)\right]\right\rangle &= \left\langle\,\exp\left[\Psi_1\left(r_1,R\right)+\Psi_2\left(r_1,R\right)+\Psi_1^*\left(r_2,R\right)\right.\right.\\ &\quad\left.\left.+\Psi_2^*\left(r_2,R\right)\right]\right\rangle\\ &= \exp\left[2E_1\left(0,0\right)+E_2\left(r_1,r_2\right)\right]\end{aligned} \tag{2.66}$$

and

$$\begin{aligned}&\left\langle\exp\left[\Psi\left(r_1,R\right)+\Psi^*\left(r_2,R\right)+\Psi\left(r_3,R\right)+\Psi^*\left(r_4,R\right)\right]\right\rangle\\ &\quad=<\ \exp\left[\Psi\left(r_1,R\right)+\Psi_2\left(r_1,R\right)+\Psi_1^*\left(r_2,R\right)+\Psi_2^*\left(r_2,R\right)+\Psi_1\left(r_3,R\right)\right.\\ &\qquad\left.+\Psi_2\left(r_3,R\right)+\Psi_1^*\left(r_4,R\right)+\Psi_2^*\left(r_4,R\right)\right]>\\ &\quad=\ \exp\left[4E_1\left(0,0\right)+E_2\left(r_1,r_2\right)+E_2\left(r_1,r_4\right)+E_2\left(r_3,r_2\right)+E_2\left(r_3,r_4\right)\right.\\ &\qquad\left.+E_3\left(r_1,r_3\right)+E_3^*\left(r_2,r_4\right)\right].\end{aligned} \tag{2.67}$$

Assuming a statistically homogeneous and isotropic random medium, $E_1\left(0,0\right)$, $E_2\left(r_1,r_2\right)$ and $E_3\left(r_1,r_2\right)$ in Eqs. (2.65), (2.66) and (2.67) are given as [29]

$$\begin{aligned}E_1\left(0,0\right) &= \langle\Psi_2\left(r,R\right)\rangle+1/2\left\langle\Psi_1^2\left(r,R\right)\right\rangle\\ &= -2\pi^2k^2\sec\left(\theta\right)\int_{h_0}^{H}\int_0^{\infty}\kappa\Phi_n\left(h,\kappa\right)d\kappa dh,\end{aligned} \tag{2.68}$$

$$\begin{aligned}E_2\left(r_1,r_2\right) &= \langle\Psi_1\left(r_1,R\right)\Psi_1^*\left(r_2,R\right)\rangle\\ &= 4\pi^2k^2\sec\left(\theta\right)\int_{h_0}^{H}\int_0^{\infty}\kappa\Phi_n\left(h,\kappa\right)\\ &\quad\cdot\exp\left(-\Lambda L\kappa^2\xi^2/k\right)\\ &\quad\cdot J_0\left[\kappa\mid\left(1-\bar{\Theta}\xi\right)\rho-2i\Lambda\xi r\mid\right]d\kappa dh\end{aligned} \tag{2.69}$$

and

$$\begin{aligned}E_3\left(r_1,r_2\right) &= \langle\Psi_1\left(r_1,R\right)\Psi_1\left(r_2,R\right)\rangle\\ &= -4\pi^2k^2\sec\left(\theta\right)\int_{h_0}^{H}\int_0^{\infty}\kappa\Phi_n\left(h,\kappa\right)\\ &\quad\cdot\exp\left(-\Lambda R\kappa^2\xi^2/k\right)J_0\left[\left(1-\bar{\Theta}\xi-i\Lambda\xi\right)\kappa\rho\right]\end{aligned}$$

$$\cdot \exp\left[-\frac{i\kappa^2 R}{k}\xi\left(1-\bar{\Theta}\xi\right)\right] d\kappa dh. \tag{2.70}$$

In the above equations, $i^2 = -1$, $\rho = r_1 - r_2$, and $r = 1/2\,(r_1 + r_2)$, $J_0\,(x)$ is the Bessel function, and $\Phi_n\,(h,\kappa)$ is the power spectrum of refractive index fluctuations commonly defined by classical Kolmogorov spectrum. The parameter ξ is the normalized distance variable and is given as $\xi = 1 - (h - h_0)\,/\,(H - h_0)$ for the uplink propagation and $\xi = (h - h_0)\,/\,(H - h_0)$ for downlink propagation. The parameter $\bar{\Theta} = 1 - \Theta$ is the complementary beam parameter. The conventional Rytov approximation is then used to calculate the mean intensity at the receiver and is given by

$$\langle I\,(r,R)\rangle = I_0\,(r,R)\,\exp\left[2E_1\,(0,0) + E_2\,(r,r)\right], \tag{2.71}$$

where I_0 is the intensity profile at the receiver without atmospheric turbulence and given by Eq. (2.58). Assuming the mean intensity to be approximated by Gaussian spatial profile, the above equation can be expressed as [30]

$$\langle I\,(r,R)\rangle = \frac{W_0^2}{W_e^2}\exp\left(\frac{-2r^2}{W_e^2}\right) \quad \left[\mathrm{W/m^2}\right], \tag{2.72}$$

where W_e is the effective spot size of the Gaussian beam in the presence of the optical turbulence. The effective spot size W_e for an uplink ground-to-satellite link is given by [16, 21]

$$W_e = W\,(1 + T_{ss})^{1/2}, \tag{2.73}$$

where

$$\begin{aligned} T_{ss} &= -2E_1\,(0,0) - E_2\,(0,0) \\ &= 4\pi^2 k^2 \sec(\theta) \int_{h_0}^{H}\int_{0}^{\infty} \kappa\Phi_n\,(h,\kappa) \\ &\quad \cdot \left\{1 - \exp\left[-\frac{\Lambda R\kappa^2}{k}\left(1 - \frac{h-h_0}{H-h_0}\right)\right]\right\} d\kappa dh. \end{aligned} \tag{2.74}$$

For downlink path, T_{ss} is given by

$$T_{ss} = 4\pi^2 k^2 \sec(\theta) \int_{h_0}^{H}\int_{0}^{\infty} \kappa\Phi_n\,(h,\kappa)$$

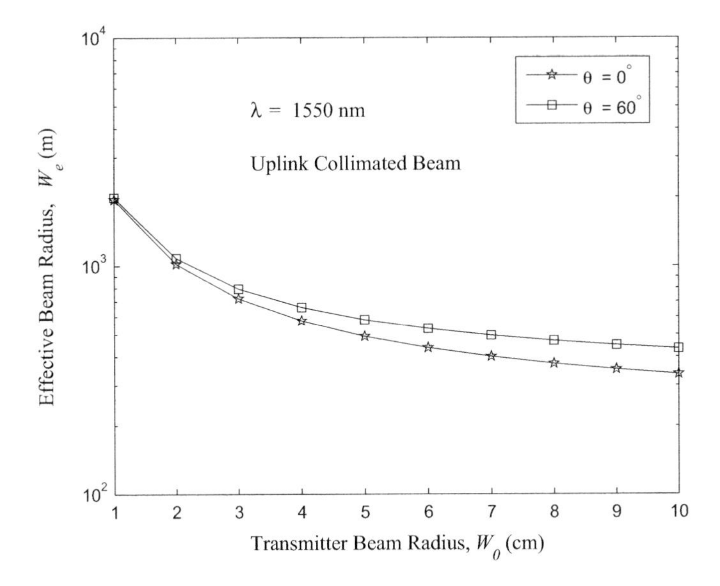

Fig. 2.17 Effective beam radius at the receiver (in m) as a function of transmitter beam radius (in cm) for various zenith angles

$$\cdot \left\{1 - \exp\left[-\frac{\Lambda R\kappa^2}{k}\left(\frac{h-h_0}{H-h_0}\right)\right]\right\} d\kappa dh. \tag{2.75}$$

It is seen from Eqs. (2.74) and (2.75) that for both uplink and downlink, turbulence-induced mean irradiance is completely determined by the beam spot size at the receiver. Figure 2.17 shows the effective beam radius/spot size at the receiver as a function of transmitter beam radius for ground-to-satellite uplink. It is clear that effective beam size at the receiver decreases rapidly up to almost 4 cm. For larger transmitter beam size, i.e., $W_0 > 4$ cm, the effective beam size decreases further, but the decrease is not very much. For downlink paths, the effective spot size in the presence of turbulence is essentially the same as the diffraction spot size W. This is due to the fact that level of turbulence at the higher altitude is less than that at the ground level.

2.1.3.2 Modified Rytov Approximation

The conventional Rytov method discussed above is generally limited to weak fluctuation conditions. So a modified version of Rytov approximation has been developed which is applicable under weak to strong atmospheric fluctuations. For this, the following basic assumptions are made:

(i) The received irradiance fluctuations can be modeled as a modulation process in which the small-scale (diffracting) and large-scale (refracting) fluctuations are multiplicative.

(ii) The small- and large-scale processes are statistically independent.
(iii) The Rytov method for optical scintillation is valid even in the saturation regime by using spatial frequency filters to account for the loss of spatial coherence of the optical wave in strong fluctuation conditions.

Therefore, the normalized irradiance is written as $I = XY$, where X and Y are the statistically independent random quantities arising from large-scale and small-scale turbulent eddies. It is assumed that X and Y have unity mean, i.e., $\langle X\rangle = \langle Y\rangle=1$. Therefore, in this case, $\langle I\rangle = 1$ and the second moment of irradiance $\langle I^2\rangle$ is given by

$$\begin{aligned}\langle I^2\rangle &= \langle X^2\rangle\langle Y^2\rangle \\ &= \left(1+\sigma_x^2\right)\left(1+\sigma_y^2\right).\end{aligned} \tag{2.76}$$

where σ_x^2 and σ_y^2 are the normalized variance of the large- and small-scale irradiance fluctuations, respectively. Further, the scintillation index is given by

$$\begin{aligned}\sigma_I^2 &= \frac{\langle I^2\rangle}{\langle I\rangle^2} - 1 \\ &= \left(1+\sigma_x^2\right)\left(1+\sigma_y^2\right) - 1 \cdot \\ &= \sigma_x^2 + \sigma_y^2 + \sigma_x^2\sigma_y^2\end{aligned} \tag{2.77}$$

Therefore, the normalized variance σ_x^2 and σ_y^2 can be written in terms of log-irradiance variance,

$$\begin{aligned}\sigma_x^2 &= \exp\left(\sigma_{\ln x}^2\right) - 1 \\ \sigma_y^2 &= \exp\left(\sigma_{\ln y}^2\right) - 1,\end{aligned} \tag{2.78}$$

where $\sigma_{\ln x}^2$ and $\sigma_{\ln y}^2$ are the large- and small-scale log-irradiance variances, respectively. Hence, total scintillation index will be

$$\sigma_I^2 = \exp\left(\sigma_{\ln I}^2\right) - 1 = \exp\left(\sigma_{\ln x}^2 + \sigma_{\ln y}^2\right) - 1. \tag{2.79}$$

In case of weak fluctuations, the scintillation index in Eq. (2.79) reduces to the limiting form [16, 31]

$$\sigma_I^2 \approx \sigma_{\ln I}^2 \approx \sigma_{\ln x}^2 + \sigma_{\ln y}^2. \tag{2.80}$$

2.2 Atmospheric Turbulent Channel Model

Beam propagation in the atmosphere can be described by the following wave equation [32–35]

$$\nabla^2 U + k^2 n^2(r)\, U = 0, \tag{2.81}$$

where emphU and k represent electric field and wave number $(2\pi/\lambda)$, respectively, ∇^2 is the Laplacian operator given as $\nabla^2 = \partial^2/\partial x^2 + \partial^2/\partial y^2 + \partial^2/\partial z^2$. The parameter n is the refractive index of medium that is generally a random function of space and is given by Eq. (2.27). When an optical beam propagates through the atmosphere, random fluctuations in air temperature and pressure produce refractive index inhomogeneities that affect the amplitude and phase of the beam. The wavefront perturbations introduced by the atmosphere can be physically described by the Kolmogorov model. The associated power spectral density for the refractive index fluctuations is defined as

$$\Phi_n(\kappa) = 0.033 C_n^2 \kappa^{-11/3} \frac{1}{L_0} \ll \kappa \ll \frac{1}{l_0}, \tag{2.82}$$

where κ is the scalar spatial frequency (in rad/m). The value of C_n^2 is essentially fixed for horizontal propagation over reasonable distance. Typically, its value ranges from $10^{-17}\,\mathrm{m}^{-2/3}$ (for weak turbulence) up to $10^{-13}\,\mathrm{m}^{-2/3}$(for strong turbulence). However, for vertical or slant propagation, the C_n^2 varies as a function of height above the ground. In that case, the average value over the entire propagation path is taken. It is dependent on various parameters like temperature, atmospheric pressure, altitude, humidity, wind speed, etc. Therefore, it is convenient to model the strength of turbulence in the atmosphere based on some empirical scintillation data.

The measured data for C_n^2 can be classified into boundary layer and free space depending upon the height above the ground. The boundary layer is the region close to the Earth surface having large temperature and pressure fluctuations resulting in large convective instabilities. This region extends from hundreds of meters to about 2 km above the surface. Further, large variations in the value of C_n^2 are observed depending upon the location, time of the day, wind speed, and solar heating. A good example of boundary layer C_n^2 measurement shows diurnal variations with peak at afternoon, dips during neutral duration close to sunrise and sunset, and almost constant value at night. The date time measurement of structure constant profile as a function of altitude shows $(-4/3)$ dependence of C_n^2 on altitude. On the other hand, the free atmosphere layer involves the altitude in the vicinity of the tropopause (15–17 km) and higher altitudes. The value of C_n^2 at higher altitudes is very small. Based on these measurements, various empirical models of C_n^2 have been proposed [36, 37]. All these models describe the strength of the atmospheric turbulence with respect to the altitude. As it is not easy to capture all the variations of C_n^2, none of the models describe the characteristic of turbulence with sufficient accuracy. The structure parameter constant C_n^2 evaluated from Fried model is shown in Fig. 2.18. This model is one of the oldest models and is given as [2]

$$C_n^2(h) = K_0 h^{-1/3} \exp\left(-\frac{h}{h_0}\right), \tag{2.83}$$

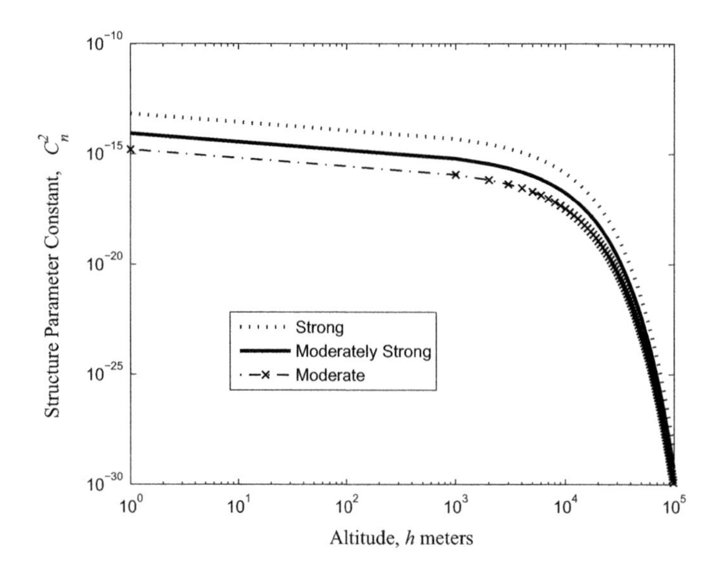

Fig. 2.18 Variations of atmospheric structure constant with altitude for the Fried model

where K_0 is the turbulence strength parameter in units of $m^{-1/3}$ and h the altitude in meters. Typical values of K_0 for strong, moderately strong, and moderate turbulence are

$$K_0 = \begin{cases} 6.7 \times 10^{-14} & \text{Strong} \\ 8.5 \times 10^{-15} & \text{Moderately strong} \\ 1.6 \times 10^{-15} & \text{Moderate} \end{cases} \tag{2.84}$$

The Hufnagel-Valley Boundary (HVB) model [20] based on various empirical scintillation data of the atmosphere is given below

$$\begin{aligned} C_n^2(h) = 0.00594 &\left[\left(\tfrac{V}{27}\right)^2 \left(10^{-5}h\right)^{10} \exp(-h/1000) \right. \\ &\left. + 2.7 \times 10^{-16} \exp\left(-\tfrac{h}{1500}\right) + A \, \exp\left(-\tfrac{h}{100}\right) \right] m^{-2/3} \end{aligned}, \tag{2.85}$$

where V^2 is the mean square value of the wind speed in m/s, h is the altitude in meters, and A is a parameter whose value can be adjusted to fit various site conditions. The parameter A is given as

$$A = 1.29 \times 10^{-12} r_0^{-5/3} \lambda^2 - 1.61 \times 10^{-13} \theta_0^{-5/3} \lambda^2 - 3.89 \times 10^{-15}. \tag{2.86}$$

In the above equation, θ_0 is the isoplanatic angle [38] (angular distance over which the atmospheric turbulence is essentially unchanged) and r_0 the atmospheric coherence length as defined earlier. For $\theta_0 = 7\,\mu$rad and $r_0 = 5$ cm at $\lambda = 1550$ nm, calculated value of A is $3.1 \times 10^{-13}\mathrm{m}^{-2/3}$. These values correspond to HVB 5/7 model. The root mean square wind velocity V between altitude of 5 and 20 km above the sea level is given by

$$V = \left[\frac{1}{15}\int_5^{20} v^2(h)\,dh\right]^{1/2}. \tag{2.87}$$

Another very simple model given by Hufnagel and Stanley (HS) [2] is

$$C_n^2(h) = \begin{cases} \frac{1.5\times10^{-13}}{h} & h \le 20\,\mathrm{km} \\ 0 & h > 20\,\mathrm{km} \end{cases}, \tag{2.88}$$

where h is the altitude above the ground and is assumed to be less than 2500 m. A new model based on empirical and experimental observations of $C_n^2(h)$ is CLEAR 1 model [20]. This model typically describes the nighttime profile of refractive index structure constant for altitude $1.23\,\mathrm{km} < h < 30\,\mathrm{km}$. It is obtained by averaging and statistically interpolating the observations obtained over a large number of meteorological conditions. It is given as [39]

$$C_n^2(h) = \begin{cases} 10^{-17.025-4.3507h+0.814h^2} & \text{for } 1.23 < h \le 2.13\,\mathrm{km} \\ 10^{-16.2897+0.0335h-0.0134h^2} & \text{for } 2.13 < h \le 10.34\,\mathrm{km} \\ 10^{-17.0577-0.0449h-0.00051h^2+0.6181\exp(-0.5(h-15.5617)/12.0173)} & \text{for } 10.34 < h \le 30\,\mathrm{km} \end{cases}, \tag{2.89}$$

where h is the altitude expressed in km. The refractive index structure constant, $C_n^2(h)$, is likely to be zero above 30 km. Another $C_n^2(h)$ model is submarine laser communication (SLC) [16] that was developed in Maui, Hawaii. The daytime profile of $C_n^2(h)$ is given by

$$C_n^2(h) = \begin{cases} 8.4\times10^{-15} & h \le 18.5\,\mathrm{m} \\ \frac{(3.13\times10^{-13})}{h} & 18.5 < h \le 240\,\mathrm{m} \\ 1.3\times10^{-15} & 240 < h \le 880\,\mathrm{m} \\ \frac{8.87\times10^{-7}}{h^3} & 880 < h \le 7200\,\mathrm{m} \\ \frac{2\times10^{-16}}{\sqrt{h}} & 7200 < h \le 20{,}000\,\mathrm{m} \end{cases}. \tag{2.90}$$

A comparison of four models, i.e., HVB, HS, CLEAR 1, and SLC models, of atmospheric profile of the refractive index structure constant is shown in Fig. 2.19. Table 2.5 gives the description of various turbulence profile models used in FSO communication systems.

It is observed from Fig. 2.18 that Fried model is applicable for short-range propagation path for various strengths of atmospheric turbulence. HS, CLEAR 1, and SLC models are applicable for long-range propagation, but they cannot provide

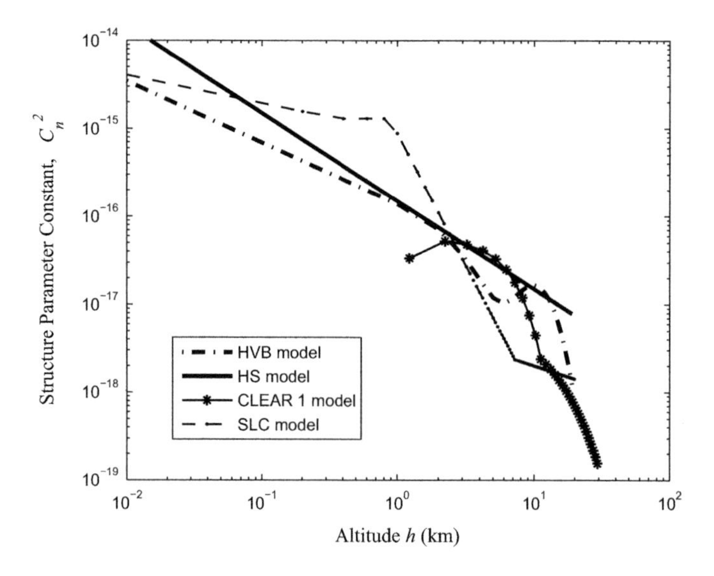

Fig. 2.19 Comparison of HVB, HS, CLEAR 1, and SLC models for atmospheric structure parameter constant

fairly good information about different site conditions. Hence, a very careful choice of $C_n^2(h)$ has to be made for determining the correct strength of turbulence in the atmosphere. The HVB model of atmospheric turbulence is most widely used for ground-to-satellite communication as it agrees fairly well with the measured values. This model includes part of upper atmosphere from 3 to 24 km, and it applies for both daytime and nighttime measurements. Further, it allows the adjustment of two parameters, i.e., coherence length, r_0, and iso-planatic angle, θ_0 in the model to simulate various site conditions.

The $C_n^2(h)$ profile as a function of height above the ground for various rms wind speeds V is shown in Fig. 2.20. It can be seen from this figure that there is a little effect of V up to a height of 1 km. Beyond this, wind speed governs the HVB profile behavior which shows a peak in the vicinity of 10 km. Therefore, for long-distance communication, wind speed effectively determines the fluctuations in irradiance of the received signal.

It can be seen that over the range $l_0 \leq r_0 \leq L_0$, variance of log-irradiance (or Rytov variance) can be written in terms of refractive index structure parameter, C_n^2 as

$$\sigma_R^2 \approx 2.24k^{7/6}\left(\sec(\theta)\right)^{11/6} \int_{h_0}^{H} C_n^2(h)\, h^{5/6} dh. \tag{2.91}$$

It is observed from Eq. (2.91) that log-irradiance variance increases with the increase in the value of C_n^2, zenith angle θ, or path length H. Substituting Eq. (2.85) into Eq. (2.91) gives the variance of log-irradiance σ_R^2 as

Table 2.5 Turbulence profile models for C_n^2

Models	Range	Comments
PAMELA model [40]	Long (few tens of kms)	– Robust model for different terrains and weather type – Sensitive to wind speed – Does not perform well over marine/overseas environment
NSLOT model[41]	Long (few tens of kms)	– More accurate model for marine propagation – Surface roughness is "hardwired" in this model – Temperature inversion, i.e., $(T_{air} - T_{sur} > 0)$, is problematic
Fried model [42]	Short (in meters)	– Support weak, strong, and moderate turbulence
Hufnagel and Stanley Model [43]	Long (few tens of kms)	– C_n^2 is proportional to h^{-1} – Not suitable for various site conditions
Hufnagel valley model [44, 45]	Long (few tens of kms)	– Most popular model as it allows easy variation of daytime and nighttime profile by varying various site parameters like wind speed, iso-planatic angle, and altitude – Best suited for ground-to-satellite uplink – HV 5/7 is a generally used to describe C_n^2 profile during daytime. HV5/7 yields a coherence length of 5 cm and isoplanatic angle of 7 μrad at 0.5 μm wavelength
Gurvich model [46]	Long (few tens of kms)	– Covers all regimes of turbulence from weak, moderate to strong – C_n^2 dependance on altitude, h, follows power law i.e., $C_n^2 \propto h^{-n}$ where n could be 4/3, 2/3, or 0 for unstable, neutral, or stable atmospheric conditions, respectively
Von Karman-Tatarski model [47, 48]	Medium (few kms)	– Make use of phase peturbations of laser beam to estimate inner and outer scale of turbulence – Sensitive to change in temperature difference
Greenwood model [49]	Long (few tens of kms)	– Nighttime turbulence model for astronomical imaging from mountain top sites
Submarine laser communication (SLC) [50] model	Long (few tens of kms)	– Well suited for daytime turbulence profile at inland sites – Developed for AMOS observatory in Maui, Hawaii
Clear 1 [20]	Long (few tens of kms)	– Well suited for nighttime turbulent profile – Averages and statistically interpolate radiosonde observation measurements obtained from large number of meteorological conditions

(continued)

Table 2.5 (continued)

Models	Range	Comments
Aeronomy laboratory model (ALM) [51]	Long (few tens of kms)	– Shows good agreement with radar measurements – Based on relationship proposed by Tatarski [48] and works well with radiosonde data
AFRL radiosonde model [52]	Long (few tens of kms)	– Similar to ALM but with simpler construction and more accurate results as two separate models are used for troposphere and stratosphere – Daytime measurements could give erroneous results due to solar heating of thermosonde probes

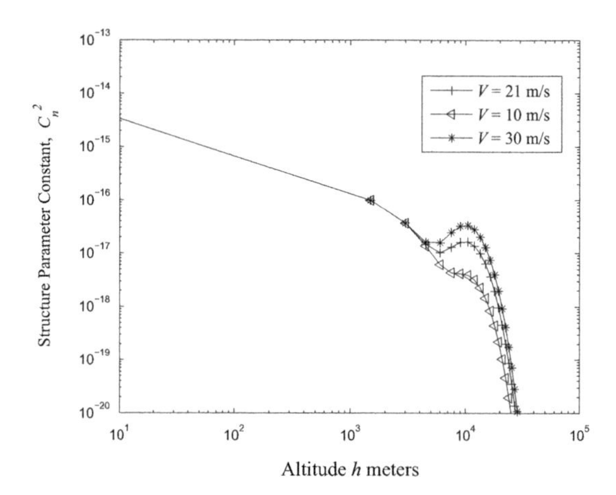

Fig. 2.20 $C_n^2\,(h)$ profile as a function of altitude

$$\sigma_R^2 \approx \left[7.41 \times 10^{-2}\left(\frac{V}{27}\right)^2 + 4.45 \times 10^{-3}\right]\lambda^{-7/6}\,(\sec(\theta))^{11/6}. \qquad (2.92)$$

Variations of σ_R^2 with rms wind velocity V for different θ at $\lambda = 1064$ nm obtained from this equation are shown in Fig. 2.21. It is observed that the irradiance fluctuations increase with the increase in V and is less at lower value of θ for the same value of V. Typical value of θ lies between 0° and 60°.

2.3 Techniques for Turbulence Mitigation

In FSO communication, turbulence in the atmosphere results in irradiance fluctuations or beam wander effect of received signal, leading to an increased bit error rate (BER) in the system. Irradiance fluctuations (or scintillation) and beam wander effect induce deep signal fades when an optical signal propagates through turbulent

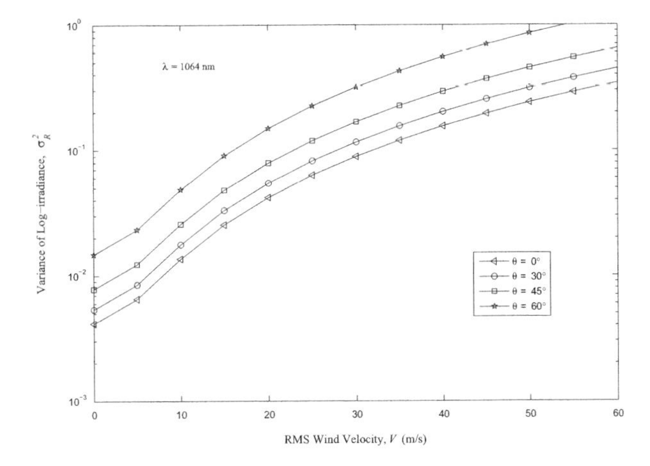

Fig. 2.21 Log-irradiance variance as a function of rms wind velocity V for zenith angle $\theta = 0°, 30°, 40°$ and $60°$

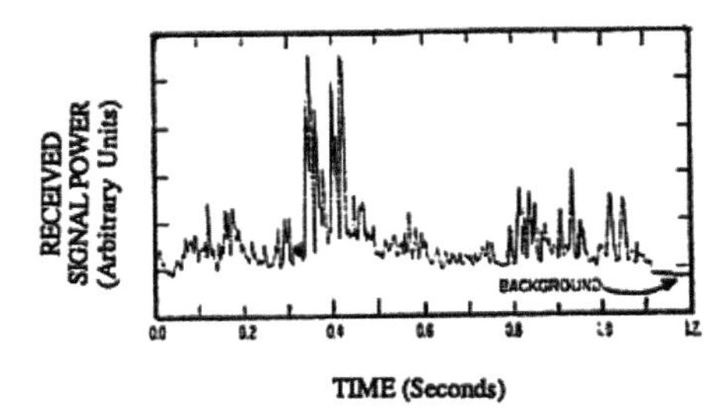

Fig. 2.22 Power fluctuations for small detector placed 145 km from the transmitter [54]

atmosphere. This deep signal fade lasts for 1–100 μs [53]. If a link is operating at say, 1 Gbps, this could result in the loss of up to 10^5 consecutive bits. This introduces the burst errors that effectively degrade the performance of the FSO link and reduce the system availability. Therefore, some mitigation technique has to be employed in order to avoid this huge data loss due to turbulence-induced irradiance fluctuations and beam wander. Various techniques to mitigate the effect of turbulence-induced signal fading are discussed below.

2.3.1 Aperture Averaging

If the size of receiver aperture is considerably smaller than the beam diameter, then the received beam will experience lots of intensity fluctuations due to turbulence in the atmosphere. Typical power fluctuations recorded by the detector of small area are shown in Fig. 2.22.

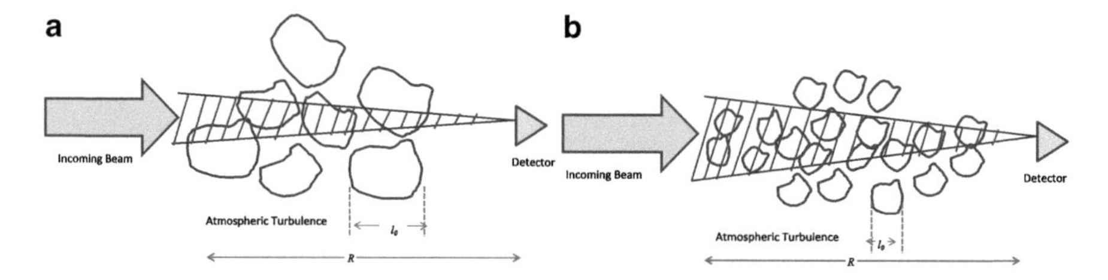

Fig. 2.23 Scattered optical signal from turbulent cells within acceptance cone (**a**) geometrical optics hold good if cone width is less than the cell dimension and (**b**) diffraction effect becomes important if cone width include many turbulent cells

The fluctuations in the received power can be explained with the help of illustration shown in Fig. 2.23. The atmosphere will provide an acceptance angle for the receiver assuming detector at the receiver to be omnidirectional. The scattered optical signal from turbulent cells within the acceptance cone will contribute to the received signal power. These turbulent cells form a diffracting aperture with average size l, and therefore, the angle formed by the acceptance cone is given by $\theta \cong \lambda/l$. The largest acceptance cone will be for smallest eddy size (inner scale of turbulent eddy), l_0, so that

$$\theta_{\max} \cong \frac{\lambda}{l_0}. \tag{2.93}$$

The maximum width of the cone is $R\theta_{\max}$, and as long as this width is less than inner scale of turbulent eddy, l_0, geometrical optics produce good results. Therefore, geometrical optics is valid when [55]

$$\theta_{\max} R = \frac{\lambda}{l_0} \cdot R < l_0, \tag{2.94}$$

or

$$\sqrt{\lambda R} < l_0. \tag{2.95}$$

However, if $\sqrt{\lambda R} \geq l_0$, then the acceptance cone may contain many smaller cells, and the received power will experience more power fluctuations as long as receiver aperture is less than beam diameter. If the diameter of the receiver aperture is made larger, receiver will average out the fluctuations over the aperture, and the irradiance fluctuations will be less than that of point receiver. This effect can be very well understood with the help of Fig. 2.24. The figure shows receiver aperture with dark and light speckles of size $\sqrt{\lambda R}$. If there is a point receiver, it will collect only one speckle that will fluctuate randomly leading to degradation in system performance. However, if the size of receiver aperture is increased, it will enhance the received power level and average out fluctuations caused by these speckles leading to

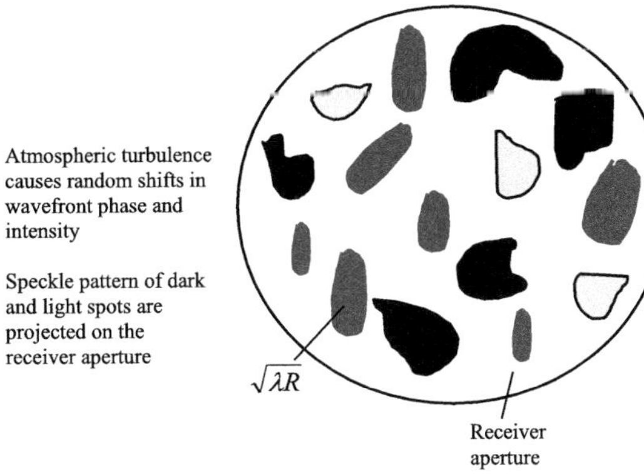

Atmospheric turbulence causes random shifts in wavefront phase and intensity

Speckle pattern of dark and light spots are projected on the receiver aperture

$\sqrt{\lambda R}$

Receiver aperture

Fig. 2.24 Speckle spot formation on the receiver plane [56]

improved BER performance. The parameter that is usually used to quantify the reduction in power fluctuations by aperture averaging is called aperture averaging factor, A_f [57]. It is defined as the ratio of normalized variance of the irradiance fluctuations from a receiver with aperture diameter D_R to that from a receiver with a point aperture, i.e.,

$$A_f = \frac{\sigma_I^2(D_R)}{\sigma_I^2(0)}, \tag{2.96}$$

where $\sigma_I^2(D_R)$ and $\sigma_I^2(0)$ are the scintillation indices for receiver with aperture diameter D_R and point receiver ($D_R \approx 0$), respectively. In [57] the factor A_f is approximated, and it is given by

$$A_f \simeq \left[1 + A_0\left(\frac{D_R^2}{\lambda h_0 \sec(\theta)}\right)^{7/6}\right]^{-1}, \tag{2.97}$$

where $A_0 \approx 1.1$ and h_0 the atmospheric turbulence aperture averaging scale height. It is given as [57]

$$h_0 = \left[\frac{\int_0^H C_n^2(h)\, h^2 dh}{\int_0^H C_n^2(h)\, h^{5/6} dh}\right]^{6/7}. \tag{2.98}$$

This relation takes into account the slant path propagation through the atmosphere and permits the modeling of the atmospheric refractive index function. Figure 2.25 shows the variations of aperture averaging factor for different aperture diameters using HVB 5/7 model with $V = 21$ m/s. It is evident from this figure that there is an improvement in the aperture averaging factor implying decrease in

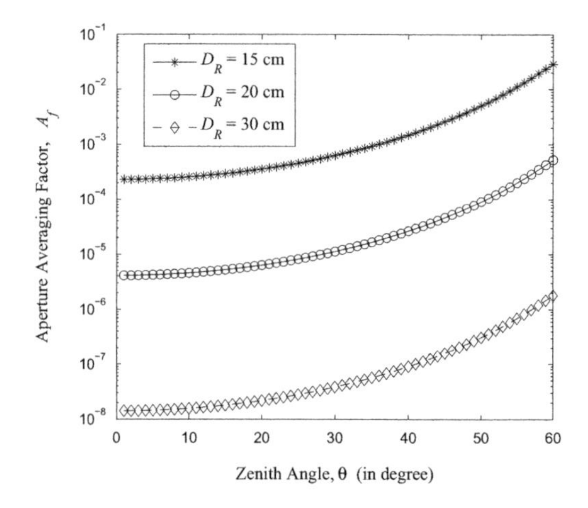

Fig. 2.25 Variations of aperture averaging factor with zenith angle θ for various values of receiver aperture diameter $D_R = 15, 20$ and 30 cm using HVB 5/7 model

the value of A_f with the increase in the receiver aperture diameter D_R. Further, the aperture averaging factor degrades with the increase in zenith angle.

2.3.2 *Spatial Diversity*

It is not always possible to increase the receiver aperture diameter beyond a certain level as it will lead to increase in background noise. Hence, increasing receiver aperture may not be an optimum solution. To achieve the same level of performance as an aperture integrator receiver, the single large aperture is replaced by array of small apertures (either at the transmitter or at the receiver) that are sufficiently separated from each other. The separation between the multiple apertures should be greater than the coherence length, r_0, of the atmosphere so that multiple beams are independent and at least uncorrelated. This technique of employing multiple apertures either at the transmitter (also called transmit diversity) or at the receiver (also called receive diversity) or at both sides (also called multiple input multiple output – MIMO) to mitigate the effect of turbulence is known as spatial diversity [58] as shown in Fig. 2.26. If a single laser beam is used for transmission through the atmosphere, turbulence in the atmosphere will cause the beam to split up into various small beam segments. These segments will then independently move around due to local changes in the refractive index of the atmosphere. At the receiver, various segments will either combine inphase or out of phase with respect to each other. Inphase events will cause surge in the power, whereas out-of-phase events will cause severe signal fades, and these lead to random fluctuations in the power at the receiver. If instead of single beam, multiple independent, and uncorrelated spatially diverse beams are used for transmission, then any overlapping of beams at the receiver will result in addition of power from different beams. Furthermore, the probability of deep fades and surges will be reduced significantly.

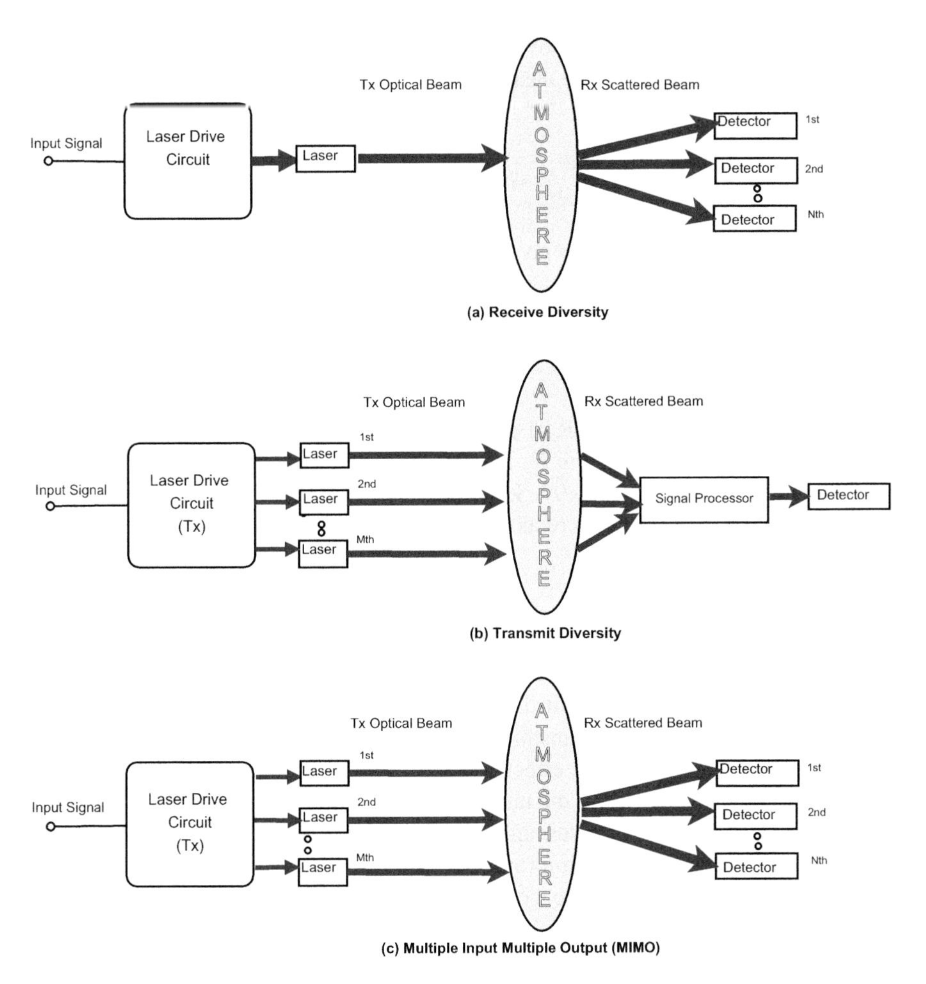

Fig. 2.26 Concept of (**a**) receive diversity, (**b**) transmit diversity and (**c**) multiple input multiple output (MIMO) techniques

Besides mitigating irradiance fluctuations due to turbulence, it also helps the FSO system designer to limit the transmitter power within allowable safe laser power limits. To determine the improvement in the performance with multiple apertures to that of single large aperture, let us take the case of array of N statistically independent detectors. In this case, the summed output is given by

$$I_r = \eta \sum_{j=1}^{N} \left(I_{s,j} + I_{n,j} \right), \tag{2.99}$$

where η is the optical to electrical conversion efficiency and $I_{s,j}$ and $I_{n,j}$ are signal and noise currents corresponding to j^{th} receiver, respectively. For simplicity, it is assumed that mean and variance of signal and noise currents in all receivers are the same. With this assumption, the mean and the variance of the total received current I_r are given as

$$\langle I_r \rangle = N \langle I_{s,1} \rangle, \ \sigma_{I_r}^2 = N\left[\langle I_{s,1}^2 \rangle - \langle I_{s,1} \rangle^2 + \langle I_{n,1}^2 \rangle\right] = N\left(\sigma_{s,1}^2 + \sigma_{n,1}^2\right). \qquad (2.100)$$

Therefore, the mean rms SNR is given by

$$\langle SNR_N \rangle = \frac{N \langle I_{s,1} \rangle}{\sqrt{N\left(\sigma_{s,1}^2 + \sigma_{n,1}^2\right)}} = \sqrt{N} \langle SNR_1 \rangle, \qquad (2.101)$$

where $\langle SNR_1 \rangle$ is the mean SNR of a single detector receiver. The above equation shows that the output SNR from N -independent detectors can improve the system performance by a factor of $\sqrt{N}$. Likewise, the effective normalized irradiance variance (scintillation index) is reduced by N i.e., [16]

$$\sigma_{I,N}^2 = \frac{1}{N}\sigma_{I,1}^2, \qquad (2.102)$$

where N is the number of multiple detectors at the receivers. This is true in case of multiple transmitters (or transmit diversity) as well wherein multiple independent and uncorrelated beams are transmitted toward the receiver. The use of multiple beams significantly reduces the effect of turbulence-induced scintillation.

The number of detectors or the number of transmit beams required to achieve a given BER depends upon the strength of atmospheric turbulence. In principle, the received irradiance statistics is improved with the increase in the number of transmit or received antennae. The practical considerations, such as system complexity, cost, efficiency of transmit laser power, modulation timing accuracy, and the availability of space, will however limit the order of diversity to less than ten.

2.3.3 *Adaptive Optics*

Adaptive optics is used to mitigate the effect of atmospheric turbulence and helps to deliver an undistorted beam through the atmosphere. Adaptive optics system is basically a closed-loop control where the beam is pre-corrected by putting the conjugate of the atmospheric turbulence before transmitting it into the atmosphere [59]. In this way, it is able to reduce the fluctuations both in space and time. An adaptive optics system consists of wavefront sensor to measure the closed-loop phase front, corrector to compensate for the phase front fluctuations, and a deformable mirror that is driven by a suitable controller. Figure 2.27 shows the conceptual block diagram of adaptive optics system. Here, the output phase front

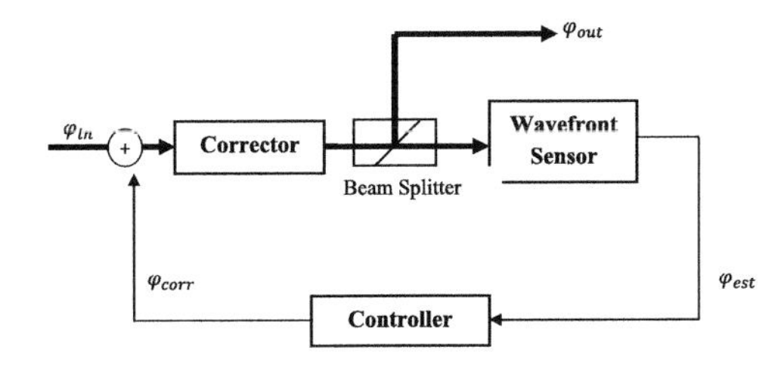

Fig. 2.27 Block diagram of an adaptive optics system

φ_{out} of the light is measured by wavefront sensor, and an estimated phase front is constructed φ_{est}. This phase information is in turn used by the controller to drive the corrector and subtract a quantity φ_{corr} from the input phase φ_{in} in order to compensate it (refer to Fig. 2.27).

The adaptive optics is mainly used for astronomical observations. But the overall objective of adaptive optics is different for optical communication relative to astronomical observations. In astronomy, the objective is to increase the sharpness of the images, and any loss of signal energy can be made up by longer observation time. But in optical communication, the signal energy for a data bit is fixed and must be conserved for efficient communication. Therefore, an adaptive optics system has to implement an optimization system to minimize the overall FOV, i.e., to minimize the amount of background light while maximizing the amount of desired signal energy captured. This makes the design of adaptive optics system complicated and results in increased system cost.

2.3.4 Coding

Error control coding can be used in FSO link to mitigate the effect of turbulence-induced scintillation. Coding usually involves the addition of extra bits to the information which later helps the receiver in correcting the errors introduced during transmission through turbulent atmosphere. Thus, depending upon the strength of turbulence and the link range, the choice of appropriate coding techniques can significantly reduce the BER of the FSO system which can be traded for reduction in the required signal power to close the link. Besides coding, interleaving can also be used especially when the channel is bursty. The channel coding not only improves information carrying capacity but also reduces the required signal power at the receiver. The reduction in the required power level is commonly referred to as coding gain and is defined as the difference in the power levels between uncoded (in dB) and coded system (in dB) to reach the same BER level and is given as

$$\Gamma_{\text{code}} = 10\log_{10}\left[\frac{P_{req}\,(\text{uncoded})}{P_{req}\,(\text{coded})}\right]. \tag{2.103}$$

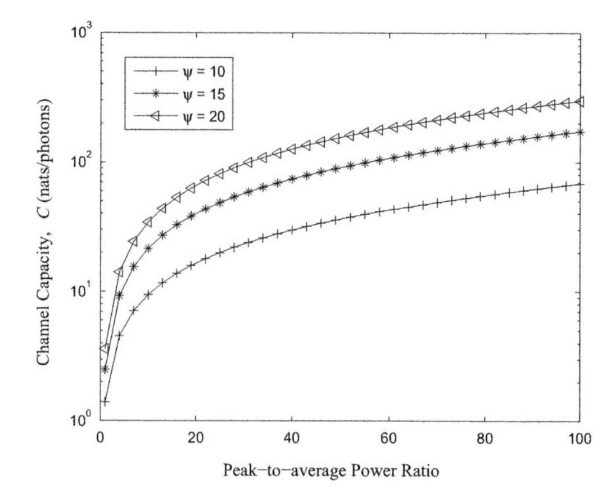

Fig. 2.28 Channel capacity vs. peak-to-average power ratio for various ratios of signal and background photon arrival rates

According to Shannon's theorem, for a given noisy channel having capacity emphChannel capacityC and information transmission rate R_b, the probability of error can be made arbitrarily small by suitable coding technique if $R_b \leq C$. This means, theoretically it is possible to transmit information without error if the rate of transmission is below the limiting channel capacity C. The channel capacity for optical channel in the presence of background noise is given as [60]

$$C = (\log_2 e)\frac{\lambda_s}{\mathbb{P}}\left[\left(1+\frac{1}{\psi}\right)\ln(1+\psi) - \left(1+\frac{\mathbb{P}}{\psi}\right)\ln\left(1+\frac{\psi}{\mathbb{P}}\right)\right], \tag{2.104}$$

where λ_s is the rate of arrival of signal photons (in photons/sec), $\psi = \lambda_s/\lambda_B$ the detected peak signal to background photon ratio, and $\mathbb{P}$ the peak-to-average power ratio of the signal. The variations of channel capacity C with peak-to-average power ratio $\mathbb{P}$ from the above equation are shown in Fig. 2.28.

One of the methods commonly used to improve the channel capacity in FSO communication is to increase peak-to-average power ratio of the signal. This ratio is very much dependent upon the choice of modulation scheme. Therefore, the channel capacity can be improved by using a modulation scheme that has high peak-to-average power ratio. For this reason, $\mathbb{M}$-PPM is one of the suitable modulation schemes for FSO communication system.

2.3.5 *Hybrid RF/FSO*

The performance of FSO communication is drastically affected by weather conditions and atmospheric turbulence. This can lead to link failures or poor BER performance of FSO system. Therefore, in order to improve the reliability and improve the availability of the link, it is wise to pair up FSO system with a more reliable RF system. Such systems are called hybrid RF/FSO and are capable of providing high link availability even in adverse weather conditions [12]. The major cause of signal degradation in RF transmission is due to rain (as the carrier wavelength is comparable to the size of the raindrop) and in FSO communication is due to fog. So, the overall system availability can be improved by using low data rate RF link as a backup when FSO link is down. In [61], the availability of an airborne hybrid RF/FSO link is evaluated. It was observed that the FSO link provides poor availability during low cloud conditions due to the attenuation by cloud particles and temporal dispersion. However, a significant improvement was observed when a hybrid RF/FSO link was used as RF signals are immune to cloud interference. The conventional approach of hybrid RF/FSO causes inefficient use of channel bandwidth [62]. Also, a continuous switching between RF and FSO system could bring down the entire system. Therefore, a new approach as suggested in [63] gives symbol rate adaptive joint coding scheme wherein both RF and FSO subsystems are active simultaneously and save channel bandwidth. Hybrid channel coding is also capable of utilizing both the links by combining nonuniform codes and rate adaptive codes where their code rates are varied according to the channel conditions [64].

Hybrid RF/FSO link provides great application in mobile ad hoc networks (MANETs) [65]. A reconfigurable networking environment can be formed in MANETs by combination of wireless sensor network (WSN) technology and mobile robotics. The performance of this network is, however, limited by the per node throughput provided by RF-based communication. Therefore, the combination of RF and FSO provides tremendous increase in per node throughput of MANETs. The implementation of hybrid RF/FSO MANET with real-time video data routing across 100 Mbps optical link and 802.11g RF transceiver has been studied in [66].

The RF wireless network poses a strong limitation on its capacity and throughput owing to growing development in communication technology [67]. With the increasing number of users, the chances of interference from the neighboring nodes increase and that limit the performance of the RF system. FSO system on the other hand is highly directional and has very narrow beam divergence. This makes FSO system immune from any kind of interference. Therefore, the combination of FSO and RF can help in solving the capacity scarcity problem in RF networks. The through capacity of hybrid RF/FSO link is given in [68–70].

2.4 Summary

As the optical signal propagates through the atmospheric channel, it encounters variation in the intensity of the signal due to various unpredictable environmental factors like fog, rain, snow, etc. Other factors responsible for degrading the quality of the optical beam in FSO communication are absorption and scattering, beam divergence loss, free-space loss, and pointing loss. Also, turbulence in the atmosphere causes random fluctuations in the intensity and phase of the received signal. Effect of atmospheric turbulence on the Gaussian beam has been analyzed by conventional and modified Rytov approximation. Atmospheric turbulent channel models have been discussed based on various empirical scintillation data of the atmosphere. Various statistical models to describe the irradiance statistics of the received signal due to randomly varying turbulent atmospheric channel, lognormal, negative exponential, gamma-gamma, etc. have been discussed. Various techniques to mitigate the channel fading due to atmospheric turbulence are described.

Bibliography

1. R.N. Clark, *Spectroscopy of Rocks and Minerals, and Principles of Spectroscopy & in Manual of Remote Sensing (Chapter 1)*, vol. 3. (Wiley, New York, 1999) (Disclaimer: This image is from a book chapter that was produced by personnel of the US Government therefore it cannot be copyrighted and is in the public domain)
2. R.M. Gagliardi, S. Karp, *Optical Communications*, 2nd edn. (Wiley, New York, 1995)
3. R.K. Long, Atmospheric attenuation of ruby lasers. Proc. IEEE **51**(5), 859–860 (1963)
4. R.M. Langer, Effects of atmospheric water vapour on near infrared transmission at sea level, in *Report on Signals Corps Contract DA-36-039-SC-723351* (J.R.M. Bege Co., Arlington, 1957)
5. A.S. Jursa, *Handbook of Geophysics and the Space Environment* (Scientific Editor, Air Force Geophysics Laboratory, Washington, DC, 1985)
6. H. Willebrand, B.S. Ghuman, *Free Space Optics: Enabling Optical Connectivity in Today's Networks* (SAMS publishing, Indianapolis, 2002)
7. M. Rouissat, A.R. Borsali, M.E. Chiak-Bled, Free space optical channel characterization and modeling with focus on algeria weather conditions. Int. J. Comput. Netw. Inf. Secur. **3**, 17–23 (2012)
8. H.C. Van de Hulst, *Light Scattering by Small Particles* (Dover publications, Inc., New York, 1981)
9. P. Kruse, L. McGlauchlin, R. McQuistan, *Elements of Infrared Technology: Generation, Transmission and Detection* (Wiley, New York, 1962)
10. I.I. Kim, B. McArthur, E. Korevaar, Comparison of laser beam propagation at 785 nm and 1550 nm in fog and haze for optical wireless communications. Proc. SPIE **4214**, 26–37 (2001)
11. M.A. Naboulsi, H. Sizun, F. de Fornel, Fog attenuation prediction for optical and infrared waves. J. SPIE Opt. Eng. **43**, 319–329 (2004)
12. I.I. Kim, E. Korevaar, Availability of free space optics (FSO) and hybrid FSO/RF systems. Lightpointe technical report. [Weblink: http://www.opticalaccess.com]
13. Z. Ghassemlooy, W.O. Popoola, Terrestrial free-space optical communications, in *Mobile and Wireless Communications Network Layer and Circuit Level Design*, ed. by S.A. Fares, F. Adachi (InTech, 2010), doi:10.5772/7698. [Weblink: http://www.intechopen.com/books/mobile-and-wireless-communications-network-layer-and-circuit-level-design/terrestrial-free-space-optical-communications]
14. W.K. Hocking, Measurement of turbulent energy dissipation rates in the middle atmosphere by radar techniques: a review. Radio Sci. **20**(6), 1403–1422 (1985)
15. R. Latteck, W. Singer, W.K. Hocking, Measurement of turbulent kinetic energy dissipation rates in the mesosphere by a 3 MHz Doppler radar. Adv. Space Res. **35**(11), 1905–1910 (2005)
16. L.C. Andrews, R.L. Phillips, *Laser Beam Propagation Through Random Medium*, 2nd edn. (SPIE Optical Engineering Press, Bellinghan, 1988)
17. H.E. Nistazakis, T.A. Tsiftsis, G.S. Tombras, Performance analysis of free-space optical communication systems over atmospheric turbulence channels. IET Commun. **3**(8), 1402–1409 (2009)
18. P.J. Titterton, Power reduction and fluctuations caused by narrow laser beam motion in the far field. Appl. Opt. **12**(2), 423–425 (1973)

19. J.H. Churnside, R.J. Lataitis, Wander of an optical beam in the turbulent atmosphere. Appl. Opt. **29**(7), 926–930 (1990)
20. R.R. Beland, Propagation through atmospheric optical turbulence, in *The Infrared and Electro-Optical Systems Handbook*, vol. 2 (SPIE Optical Engineering Press, Bellinghan, 1993)
21. H. Hemmati, *Near-Earth Laser Communications* (CRC Press/Taylor & Francis Group, Boca Raton, 2009)
22. L.C. Andrews, R.L. Phillips, R.J. Sasiela, R.R. Parenti, Strehl ratio and scintillation theory for uplink Gaussian-beam waves: beam wander effects. Opt. Eng. **45**(7), 076001-1–076001-12 (2006)
23. H.T. Yura, W.G. McKinley, Optical scintillation statistics for IR ground-to-space laser communication systems. Appl. Opt. **22**(21), 3353–3358 (1983)
24. J. Parikh, V.K. Jain, Study on statistical models of atmospheric channel for FSO communication link, in *Nirma University International Conference on Engineering*-(NUiCONE), Ahmedabad (2011), pp. 1–7
25. H.G. Sandalidis, Performance analysis of a laser ground-station-to-satellite link with modulated gamma-distributed irradiance fluctuations. J. Opt. Commun. Netw. **2**(11), 938–943 (2010)
26. J. Park, E. Lee, G. Yoon, Average bit-error rate of the Alamouti scheme in gamma-gamma fading channels. IEEE Photonics Technol. Lett. **23**(4), 269–271 (2011)
27. M.A. Kashani, M. Uysal, M. Kavehrad, *A Novel Statistical Channel Model for Turbulence-Induced Fading in Free-Space Optical Systems*. PhD thesis, Cornell University, 2015
28. A.K. Ghatak, K. Thyagarajan, *Optical Electronics* (Cambridge University Press, Cambridge, 2006)
29. L.C. Andrews, W.B. Miller, Single-pass and double-pass propagation through complex paraxial optical systems. J. Opt. Soc. Am. **12**(1), 137–150 (1995)
30. L.C. Andrews, R.L. Phillips, P.T. Yu, Optical scintillation and fade statistics for a satellite-communication system. Appl. Opt. **34**(33), 7742–7751 (1995)
31. H. Guo, B. Luo, Y. Ren, S. Zhao, A. Dang, Influence of beam wander on uplink of ground-to-satellite laser communication and optimization for transmitter beam radius. Opt. Lett. **35**(12), 1977–1979 (2010)
32. N.G. Van Kampen, Stochastic differential equations. Phys. Rep. (Sect. C Phys. Lett.) **24**(3), 171–228 (1976)
33. B.J. Uscinski, *The Elements of Wave Propagation in Random Media* (McGraw-Hill, New York, 1977)
34. H.T. Yura, S.G. Hanson, Second-order statistics for wave propagation through complex optical systems. J. Opt. Soc. Am. A **6**(4), 564–575 (1989)
35. S.M. Rytov, Y.A. Kravtsov, V.I. Tatarskii, *Wave Propagation Through Random Media*, vol. 4 (Springer, Berlin, 1989)
36. N.S. Kopeika, A. Zilberman, Y. Sorani, Measured profiles of aerosols and turbulence for elevations of 2–20 km and consequences on widening of laser beams. Proc. SPIE Opt. Pulse Beam Propag. III **4271**(43), 43–51 (2001)
37. A. Zilberman, N.S. Kopeika, Y. Sorani, Laser beam widening as a function of elevation in the atmosphere for horizontal propagation. Proc. SPIE Laser Weapons Tech. II **4376**(177), 177–188 (2001)
38. G.C. Valley, Isoplanatic degradation of tilt correction and short-term imaging systems. Appl. Opt. **19**(4), 574–577 (1980)
39. D.H. Tofsted, S.G. O'Brien, G.T. Vaucher, An atmospheric turbulence profile model for use in army wargaming applications I. Technical report ARL-TR-3748, US Army Research Laboratory (2006)
40. E. Oh, J. Ricklin, F. Eaton, C. Gilbreath, S. Doss-Hammel, C. Moore, J. Murphy, Y. Han Oh, M. Stell, Estimating atmospheric turbulene using the PAMELA model. Proc. SPIE Free Space Laser Commun. IV **5550**, 256–266 (2004)
41. S. Doss-Hammel, E. Oh, J. Ricklinc, F. Eatond, C. Gilbreath, D. Tsintikidis, A comparison of optical turbulence models. Proc. SPIE Free Space Laser Commun. IV **5550**, 236–246 (2004)

42. S. Karp, R.M. Gagliardi, S.E. Moran, L.B. Stotts, *Optical Channels: Fibers, Clouds, Water, and the Atmosphere*. (Plenum Press, New York/London, 1988)
43. R.E. Hufnagel, N.R. Stanley, Modulation transfer function associated with image transmission through turbulence media. J. Opt. Soc. Am. **54**(52), 52–62 (1964)
44. R.K. Tyson, Adaptive optics and ground to space laser communication. Appl. Opt. **35**(19), 3640–3646 (1996)
45. R.E. Hugnagel, Variation of atmospheric turbulence, in *Digest of Topical Meeting on Optical Propagation Through Turbulence* (Optical Society of America, Washington, DC, 1974), p. WA1
46. A.S. Gurvich, A.I. Kon, V.L. Mironov, S.S. Khmelevtsov, *Laser Radiation in Turbulent Atmosphere* (Nauka Press, Moscow, 1976)
47. M.R. Chatterjee, F.H.A. Mohamed, Modeling of power spectral density of modified von Karman atmospheric phase turbulence and acousto-optic chaos using scattered intensity profiles over discrete time intervals. Proc. SPIE Laser Commun. Prop. Atmosp. Oce. III **9224**, 922404-1–922404-16 (2014)
48. V.I. Tatarskii, *The Effects of the Turbulent Atmosphere on Wave Propagation* (Israel Program for Scientific Translations, Jerusalem, 1971)
49. M.C. Roggermann, B.M. Welsh, *Imaging Through Turbulence* (CRC Press, Boca Raton, 1996)
50. H. Hemmati (ed.), *Near-Earth Laser Communications* (CRC Press, Boca Raton, 2009)
51. T.E. Van Zandt, K.S. Gage, J.M. Warnock, An improve model for the calculation of profiles of wind, temperature and humidity, in *Twentieth Conference on Radar Meteorology* (American Meteorological Society, Boston, 1981), pp. 129–135
52. E.M. Dewan, R.E. Good, R. Beland, J. Brown, A model for C_n^2 (optical turbulence) profiles using radiosonde data. Environmental Research Paper-PL-TR-93-2043 1121, Phillips Laboratory, Hanscom, Airforce base (1993)
53. E.J. Lee, V.W.S. Chan, Optical communications over the clear turbulent atmospheric channel using diversity: part 1. IEEE J. Sel. Areas Commun. **22**(9), 1896–1906 (2004)
54. A.L. Buck, Effects of the atmosphere on laser beam propagation. Appl. Opt. **6**(4), 703–708 (1967)
55. H. Weichel, *Laser Beam Propagation in the Atmosphere* (SPIE Press, Washington, DC, 1990)
56. S. Bloom, The physics of free space optics. Technical report, AirFiber, Inc. (2002)
57. D.L. Fried, Aperture averaging of scintillation. J. Opt. Soc. Am. **57**(2), 169–172 (1967)
58. T.A. Tsiftsis, H.G. Sandalidis, G.K. Karagiannidis, M. Uysal, Optical wireless links with spatial diversity over strong atmospheric turbulence channels. IEEE Trans. Wirel. Commun. **8**(2), 951–957 (2009)
59. S.M. Navidpour, M. Uysal, M. Kavehrad, BER performance of free-space optical transmission with spatial diversity. IEEE Trans. Wirel. Commun. **6**(8), 2813–2819 (2007)
60. A.D. Wyner, Capacity and error exponent for the direct detection photon channel – part 1. IEEE Trans. Inf. Theory **34**(6), 1449–1461 (1988)
61. W. Haiping, M. Kavehrad, Availability evaluation of ground-to-air hybrid FSO/RF links. J. Wirel. Inf. Netw. (Springer) **14**(1), 33–45 (2007)
62. H. Moradi, M. Falahpour, H.H. Refai, P.G. LoPresti, M. Atiquzzaman, On the capacity of hybrid FSO/RF links, in *Proceedings of IEEE, Globecom* (2010)
63. Y. Tang, M. Brandt-Pearce, S. Wilson, Adaptive coding and modulation for hybrid FSO/RF systems, in *Proceeding of IEEE, 43rd Asilomar Conference on Signal, System and Computers*, Pacific Grove (2009)
64. E. Ali, V. Sharma, P. Hossein, Hybrid channel codes for efficient FSO/RF communication systems. IEEE. Trans. Commun. **58**(10), 2926–2938 (2010)
65. D.K. Kumar, Y.S.S.R. Murthy, G.V. Rao, Hybrid cluster based routing protocol for free-space optical mobile ad hoc networks (FSO/RF MANET), in *Proceedings of the International Conference on Frontiers of Intelligent Computing*, vol. 199 (Springer, Berlin/Heidelberg, 2013), pp. 613–620
66. J. Derenick, C. Thorne, J. Spletzer, Hybrid Free-space Optics/Radio Frequency (FSO/RF) networks for mobile robot teams, in *Multi-Robot Systems: From Swarms to Intelligent Automata*, ed. by A.C. Schultz, L.E. Parke (Springer, 2005)

67. S. Chia, M. Gasparroni, P. Brick, The next challenge for cellular networks: backhaul. Proc. IEEE Microw. Mag. **10**(5), 54–66 (2009)
68. C. Milner, S.D. Davis, Hybrid free space optical/RF networks for tactical operations, in *Military Communications Conference (MILCOM)*, Monterey (2004)
69. A. Kashyap, M. Shayman, Routing and traffic engineering in hybrid RF/FSO networks, in *IEEE International Conference on Communications* (2005)
70. B. Liu, Z. Liu, D. Towsley, On the capacity of hybrid wireless network, in *IEEE INFOCOM'03* (2003)

Chapter 3
FSO System Modules and Design Issues

The basic functional components of FSO communication system consists of (i) optical power source, transmitter; (ii) modulation and encoding of light, modulator and encoder; (iii) acquisition, tracking, and pointing (ATP) system, fine pointing mirrors and steering optics; (iv) background suppression, filter; (v) optical transmit and receive aperture; and (vi) detector, demodulator, and decoder, receiver.

Various components of the ground-based transceiver, intervening optical channel, and onboard optical flight transceiver are shown in Fig. 3.1. It consists of transmitter, receiver, ATP system, and atmospheric channel. Before transmitting the message/data signal, a beacon signal is transmitted from ground station to satellite, and the link is established using ATP system. The uplink beacon signal is first acquired by the distant target initiating the process of acquisition. During acquisition, one station (say ground-based earth transceiver) interrogates the uncertainty area of other station (say onboard satellite) either through beam decollimation or scanning of the transmitter beam. Once the onboard satellite acquires the beacon from ground-based earth transceiver, the transition from acquisition to tracking begins. For narrow beam system, point ahead compensators are used to compensate for the propagation delay over the link range. Initially for acquiring the signal, the image of ground-based earth transceiver is captured on focal pixel array, and the centroid is computed. The ground-based position vector is determined from the computed centroid. The difference between the current onboard laser position and the ground-based position with the addition of point ahead angle (in order to account for two way travel of light between ground-based earth transceiver and onboard satellite) is the position vector that drives the beam steering mirror. Once the link is acquired and a line of sight connection is established between two stations, data communication can start. Data transfer is performed using variety of coherent and noncoherent modulation schemes and encoding techniques. For both beacon and data transfer, the signal has to pass through atmospheric optical channel which in addition to background noise introduces signal attenuation and scintillation. Therefore, sufficient optical power, pointing accuracy, and directivity are needed to

H. Kaushal et al., *Free Space Optical Communication*, Optical Networks,
DOI 10.1007/978-81-322-3691-7_3

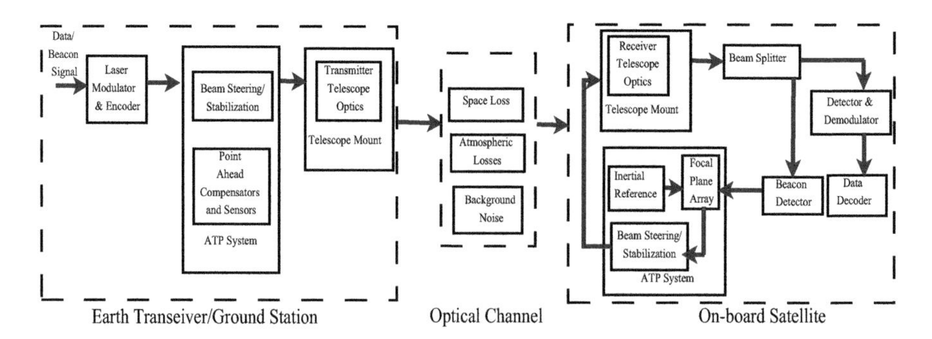

Fig. 3.1 Schematic representation of various components for ground-to-satellite optical link

deliver the required signal for uplink beacon pointing or data communication. The details of the various components used in FSO communication system are given in the following sections.

3.1 Optical Transmitter

The optical transmitter including the choice of laser, concept of ATP system, and various types of modulation schemes and coding techniques used in FSO communication system are discussed in this section. Further, the details of communication and beacon detectors in FSO receiver are also discussed. The transmitter converts the source information into optical signals which are transmitted to the receiver through the atmosphere. The essential components of the transmitter are (i) modulator, (ii) driver circuit for the optical source to stabilize the optical radiations against temperature fluctuations, and (iii) collimator that collects, collimates, and direct the optical signals toward the receiver via atmospheric channel. The optical sources that are used for FSO transmission lie in the atmospheric transmission window that is ranging from 700 to 10,000 nm wavelength. The wavelength range from 780 to 1064 nm is most widely used as beacon operating wavelength. The 1550 nm wavelength is commonly used as data operating wavelength due to following reasons:

(i) Reduced background noise and Rayleigh scattering: The absorption coefficient of the Rayleigh scattering has functional dependence with the wavelength λ as λ^{-4}. Consequently, there is almost negligible attenuation at higher operating wavelengths as compared to those at the visible range.
(ii) High transmitter power: At 1550 nm a much higher power level (almost 50 times) than at lower wavelengths is available to overcome various losses due to attenuation.
(iii) Eye-safe wavelength: The maximum permissible exposure (MPE) for eye is much higher at 1550 nm wavelength than at 850 nm. This difference can be

explained by the fact that at 850 nm, approximately 50 % of the signal can reach the retina whereas at 1550 nm, the signal is almost completely absorbed by cornea itself. And therefore the signal received at the retina is negligibly small.

The component cost increases with the increase in the operating wavelength. For good optical transmitter, the choice of laser power and wavelength has to be made very carefully so that an appropriate optical power and transmit antenna gain can be achieved in order to form a closed loop communication link. However, this is not the only constraint for most laser sources. The selection of laser is influenced by several other factors including efficiency, operational lifetime, and achievable diffraction-limited output power and weight. A good source will have narrow, stable spectral line width and nearly diffraction-limited single mode spatial profile. Some of the key requirements which affect the choice of the transmitter laser for FSO-based applications are given below:

(i) Pulse repetition frequency (PRF): The laser pulsing mechanism (e.g., Q-switching, cavity dumping) determines the PRF of the laser. Q-switched lasers using acousto-optic or electrooptic modulators have PRF less than 200 kHz. Cavity dumping lasers have PRF in the order of tens of megahertz. PRF up to several gigahertz can be achieved with the power amplified lasers used in conjunction with several stages of amplification.
(ii) Average output power: The laser should have sufficient average power for a reliable communication link with adequate link margin. For any good laser, it should provide pulse to pulse power stability and nearly constant average power over different data rates. The peak power of any laser is given by the product of energy per pulse and the pulse width. Solid-state lasers provide large peak power at low PRF. However, the maximum peak power is limited by the heat dissipation and laser safety norms.
(iii) Pulse width: Laser pulse width should be small to facilitate less background noise in narrow temporal slots.
(iv) Pulse extinction ratio: The ratio of laser power in on-mode to that of in off-mode is called pulse extinction ratio. The extinction ratio should be as large as possible. If the laser emission is not switched to complete off-mode, it may degrade the extinction ratio resulting in a lower link margin. Solid-state lasers have modulation extinction ratio of 40 dB, whereas semiconductor lasers have relatively poor extinction ratio of about 10 dB. Fiber lasers and amplifiers have extinction ratios in the order of 30 dB.
(v) Output beam quality: The output of the laser should consist of single spatial mode or at least have single null in the center of far field pattern. To avoid undesired oscillations either within the laser or in the transmitted beam, feedback isolation of the laser from the back reflected beam is required.
(vi) Beam pointing stability: For FSO-based applications, the pointing accuracy on the order of micro-radian or better is desirable. Such an accuracy requires the pointing stability of the laser to be maintained by the use of optomechanical or spatial resonators within the laser.

(vii) Overall efficiency: In order to minimize the electrical power requirement, it is desirable to have highest possible overall efficiency.
(viii) Mass and size: For any space-based applications, the mass and size of all the components should be minimized to achieve low launch cost. It therefore necessitates the use of optomechanical designs of laser resonator.
(ix) Operational lifetime: The lifetime of the active laser components (e.g., diode laser, modulator and drivers, etc.) are expected to exceed the operational lifetime of the system. Redundancy of the active elements or block redundancy of the laser will help in extending the operational lifetime. It should be noted that the higher the pump power, the lower is the expected lifetime of the laser.
(x) Thermal control and management: An efficient thermal control is required so that the dissipated heat does not affect the optical alignment integrity of the system that would otherwise result in further loss.

3.1.1 Choice of Laser

The choice of laser is driven from the basic requirements, i.e., high electrical-to-optical conversion efficiency, excellent beam quality, variable repetition rate, stable operation over lifetime, quick start up operation, and high reliability. In the early days, gas lasers such as carbon dioxide (CO_2) laser were used because they were stable and less sensitive to atmospheric effects. However, they are not very popular in FSO-based applications because of their bulky size and unreliability. Later, solid-state lasers became the choice of FSO-based applications – the most common being the neodymium/yttrium aluminum garnet (Nd/YAG). The fundamental lasing wavelength of Nd/YAG laser is at 1064 nm, and it can be doubled to 532 nm using nonlinear crystals. Other sources emitting close to Nd/YAG device include neodymiu/yttrium aluminum phosphate (Nd/YAP) and neodymium/yttrium lithium fluoride (Nd/YLF). The solid-state devices have stable and narrow spectral linewidth, and they can be designed in a pulse or continuous mode configuration. They have very high peak power (in the order of kilowatt or more) and can be operated with very narrow spectral line width (< in the order of ns).

Semiconductor laser diode, e.g., gallium arsenide (GaAs), gallium aluminum arsenide (GaAlAs), indium gallium arsenide (InGaAs), and indium gallium arsenide phosphate (InGaAsP), devices can be used for some specific FSO-based applications. Another kinds of semiconductor lasers are vertical cavity surface emitting laser (VCSEL), Fabry Perot laser, and distributed feedback laser. The threshold current requirement in VCSEL is quite low, and it allows high intrinsic modulation bandwidths in these lasers. Fabry Perot and distributed feedback (DFB) lasers have higher power density ($\approx 100\,\mathrm{mW/cm^2}$) and are compatible with EDFA. So these lasers find wide applications in FSO system. Semiconductor lasers exhibit single frequency and one spatial mode. They are also compact in size, light weight, and easy to operate. However, their output power is quite low, and therefore they require additional amplifier for long distance communication. These diodes require great care in drive electronics, otherwise they will get damaged easily. Due to reliability issues, such diodes require alternate or redundant laser source to allow smooth functioning of the system. Another kinds of lasers are erbium-doped fiber laser operating in the range of 965 to 1550 nm and master oscillator power amplifier (MOPA)-based lasers. Depending upon the amplifier architecture, these lasers can generate broadband or narrow linewidth output beams. The MOPA laser allows the amplification of oscillator through a suitable and efficient amplification medium. The oscillator and amplifier can then be individually tailored for high speed and more power, respectively. The factors that may limit the usefulness of MOPA laser is high nonlinear gain and low damage threshold for high power pulses. Fiber-based amplifiers allow tens of kilowatt of peak power and have the advantage of ease of use, efficient coupling to fibers, and relatively low noise power. However, nonlinear effects like stimulated Brillouin scattering, stimulated Raman scattering, self-phase modulation, cross-phase modulation, and four-wave mixing result in a lower SNR due to both signal reduction and introduction of additional noise.

Among existing lasers, MOPA and solid-state lasers satisfy the requirements for space-based applications. Other lasers like semiconductor lasers and EDFA are useful in multi-gigabit links for near Earth laser communication. However, fiber amplifiers have less peak power. While choosing a laser, a complex trade-off has to be made between laser power, spectral width, output wavelength, range, optical background noise, data rate, and modulation schemes to be used. Table 3.1 summarizes the various types of lasers used for FSO applications. The choice of operating wavelength for the laser has to be made vis-a-vis availability of

Table 3.1 Attributes for lasers used in FSO [6]

Laser type	Material	Wavelength (nm)	Data rate	Peak power
Solid-state pulsed	Nd/YAG	1064	<10 Mbps	Very high
	Nd/YLF	1047 or 1053		10–100 W
	ND/YAP	1080		
Solid-state mode locked	Nd/YAG	1064	Up to 1 Gbps	High
	Nd/YLF	1047 or 1053		10s of Watt
	ND/YAP	1080		
Solid-state CW	Nd/YAG	1064	>Gbps	1–5 W
	Nd/YLF	1047 or 1053		
	ND/YAP	1080		
Semiconductor pulsed	GaAlAs	780–890	1–2 Gbps	200 mW
	InGaAs	890–980	1–2 Gbps	1 W (MOPA)
	InGaAsP	1300 and	Multi-Gbps	<50 mW
	VCSEL	1550	10 Gbps	<30 mW
	Fabry Perot	780–850	40 Gbps	200 mW
	DFB	1300 and 1500		
Doped fiber amplifier	Erbium DFA	1550	10s of Gbps	1000s of Watt

components, attenuation, background noise power, and detector sensitivity at that wavelength. Most of the FSO systems are designed to operate in the 780–850 and 1520–1600 nm spectral windows due to low attenuation in these regions. Out of these, 1550 nm is most widely used for data transmission because of various factors, viz., eye safety, reduced solar background, and scattering. Further, at 1550 nm, more power can be transmitted to overcome attenuation due to fog, smog, clouds, etc. However, it suffers from certain challenges due to tight alignment requirement and higher component cost.

3.1.2 Modulators

Modulators help in converting low-frequency information signal onto optical carrier for long distance communication. The parameters that characterize optical modulators are bandwidth, insertion loss, modulation depth, drive power, and maximum optical throughput power [4]. Modulation can be applied to any one parameter, i.e., intensity, phase, frequency, or the polarization of the optical carrier. The most commonly used modulation is intensity modulation (IM) in direct detection where the intensity of the optical source is modulated in accordance with the source information. This can be achieved either by varying the direct current of the optical source or by making use of an external modulator. The use of external modulator allows high data rate transmission. However, external modulator has major disadvantages of nonlinear response, complexity, and cost. In case of coherent detection, the commonly used modulation schemes are intensity and phase modulation schemes.

A phase modulation produces frequency sidebands on a continuous wave (CW) optical beam. The amplitude of the sidebands with respect to the carrier is determined by the amplitude of the applied voltage and are given by Bessel functions. Here, the signal is applied as a voltage across the electrodes placed on top and bottom of the electrooptic crystal. This will develop electric field across the crystal. Polarized light when allowed to pass through the crystal will experience index of refraction change that will lead to change in the optical path length which is proportional to applied electric field. Therefore, the phase of the optical beam moving out of the crystal will experience phase change due to applied electric field. Figure 3.2 shows electrooptic phase modulator. This kind of modulators uses lithium niobate ($LiNbO_3$) and magnesium oxide-doped lithium niobate ($MgO/LiNbO_3$) which have high electrooptic coefficients that minimizes the required drive voltages and have high quality crystal. Also, such modulators have low insertion loss of 4.5 dB and large throughput power of more than 500 mW. The modulation bandwidth of 2 and 3 GHz can be achieved with a simple electrode structure which can be increased further by using more advanced electrode structures.

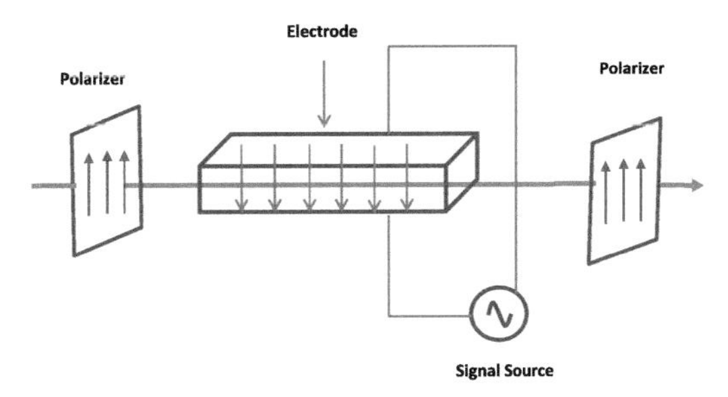

Fig. 3.2 Schematic diagram of phase modulator

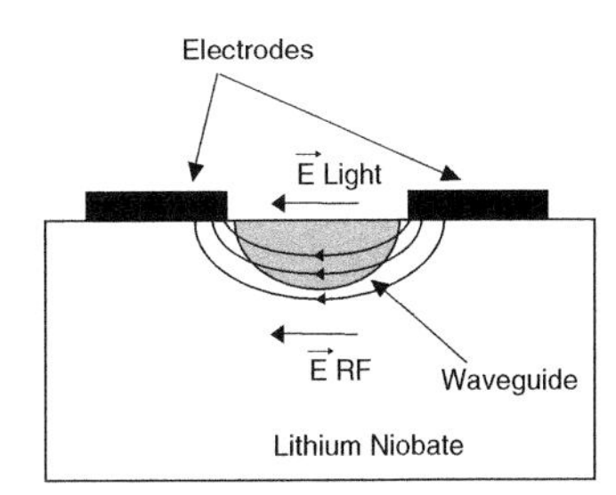

Fig. 3.3 Integrated optic $LiNbO_3$ phase modulator [9]

The phase-modulating electrooptic modulator can be efficiently used as an amplitude modulator. Mach-Zehnder modulator is one of that kind. It consists of Y-splitter which splits the incoming light into two branches. One of the branches is having phase modulator. The output light from both the branches are combined using another Y-adder. The phase of the incoming light is varied in accordance with the modulating signal. The light from both the branches will then combine either destructively or constructively depending upon the phase shift being introduced in one of the branch. This therefore controls the amplitude or the intensity of the incoming light resulting in amplitude-modulated output wave.

Integrated optical modulators (IOM) are also good choice for FSO communication and are available over broad range of wavelengths. These modulators are made from $LiNbO_3$ and are constructed using simple dielectric optical waveguide. Figure 3.3 shows the concept of integrated phase modulator. The device consists of waveguide on $LiNbO_3$ chip and electrode system. In the presence of electric field, the change in the travel time dt of light wave is given by

$$dt = dn \cdot L/c, \tag{3.1}$$

where dn is the absolute change in the index of refraction due to applied electric field, L the interaction length, and c the speed of light in vacuum. This propagation delay is equivalent to shift in phase of the output light which is given by

$$d\phi = \omega \cdot dt = \omega \cdot dn \cdot L/c, \tag{3.2}$$

where ω is the angular frequency.

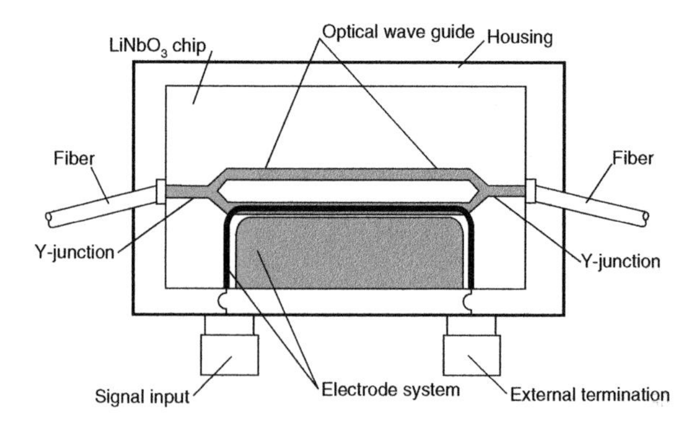

Fig. 3.4 Mach-Zehnder amplitude modulator

Similarly, integrated optics amplitude modulators constructed by patterning a Mach-Zehnder interferometer on a $LiNbO_3$ substrate is shown in Fig. 3.4. For an ideal amplitude modulator, the total optical power at the output of the modulator is given by

$$P_o = \frac{1}{2} P_i \left[1 + \cos\left(d\phi\right)\right], \tag{3.3}$$

where P_i is the input optical power and $d\phi$ the phase difference between two paths. Besides integrated optic phase and amplitude modulators, there are modulators with combinations of both phase and amplitude modulators on the same chip. The design and construction of these modulators are complex than simple IOM modulators; however, their basic principle of operation is same as that of linear electrooptic modulators.

3.1.2.1 Modulation Schemes

There are variety of modulation schemes that are suitable for FSO communication system. Both binary and multi-level modulation signaling schemes can be used in conventional FSO communication system. However, most popularly used modulation scheme is based on binary signaling for which system design is simple and inexpensive. In binary signaling scheme, the information is transmitted in each symbol period by variation of two intensity levels. On-off keying (OOK) and binary pulse-position modulation (PPM) are the most popularly used binary signaling modulation schemes in FSO communication. These modulation schemes have been discussed in Chap. 1.

The variants of PPM scheme which are becoming popular these days are differential PPM (DPPM), differential amplitude PPM (DAPPM), overlapping PPM (OPPM) [8], and combinatorial PPM (CPPM) [1]. All these modulation schemes are obtained by simple modification of PPM to achieve improved power and bandwidth efficiency. In DPPM, the empty slots following the pulse in PPM symbol are removed therefore reducing the average symbol length and improving the bandwidth efficiency. For long sequence of zeros, there could be a problem of slot synchronization; however, this problem can be overcome by making use of guard band/slot immediately after the pulse has been removed [3]. The DAPPM is a combination of DPPM and pulse amplitude modulation (PAM). It is therefore a multi-level modulation scheme where the symbol length varies from $1, 2, \ldots, \mathbb{M}$ and the pulse amplitude is selected from $1, 2, \ldots, A$ levels.

Before transmitting the optical-modulated signal, its power level has to be raised in order to compensate the huge loss in the atmospheric channel. For this purpose, optical amplifier can be used in FSO communication system as they enable the direct amplification of light with a minimum of electronics. An optical post amplifier can boost the output power of a transmitter by about 15–20 dB. Further, at front end of the receiver, the received optical signal may be very weak. The usage of optical preamplifier before the photodetector may improve the receiver sensitivity by about 10 dB. The optical post and preamplifiers are briefly discussed in Sect. 3.3.

3.2 Optical Receiver

The receiver helps to recover the transmitted data after propagating through turbulent atmosphere. It consists of receiver telescope, filter, photodetector, signal processing unit, and demodulator. The receiver telescope comprises of lenses that focus the received optical signal onto the photodetector. The filter is used to reduce the amount of background noise. The noise sources present at the receiver include background, detector dark current, preamplifier, signal shot noise, and thermal noise. The photodetector converts the received optical signal into electrical signal which is passed on to the processing unit and then to the demodulator. In the receiver, both PIN and APD can be used. In the FSO uplink, the received power level is quite low due to large free-space loss. At this power level, an APD receiver gives much better performance than the PIN receiver.

The choice of optical receiver depends upon various fundamental issues and hardware parameters. Some of the important parameters are listed below:

(i) Modulation technique: The detection technique used at the receiver depends upon the modulation format. Not every detection technique is suited for every modulation format, e.g., direct detection receivers are insensitive to phase and polarization information.
(ii) Hardware availability, reliability, and cost: Different types of receiver have different hardware requirements which may or may not be readily available at reasonable cost. For example, high gain Si-APD works efficiently only at wavelength below 1000 nm. At higher wavelengths, other detectors like InGaAs/InGaAsP can be used depending upon the requirement.

(iii) Receiver sensitivity: This is a very important parameter in all optical communication system including FSO communication system. It is measured in terms of average received photons per bit and is given as

$$n_{av} = \frac{P_{av}}{h\nu R}, \tag{3.4}$$

where $h\nu$ is the photon energy at the transmit operating wavelength ($\lambda = c/\nu$) and R_b the bit rate/data rate. The receiver sensitivity depends mainly on photon detection technique, modulation format, photodetector, and background noise. These are briefly described below:

(a) Photon detection technique: As mentioned earlier, the detection technique used in the optical receiver can be broadly classified into two types, viz., coherent detection and noncoherent direct detection. In coherent detection receiver, incoming signal is mixed with the strong local oscillator (LO) signal. This mixing of weak signal and strong LO signal at the front end of the receiver provides linear amplification and converts the optical signal into electrical signal. The strong field of LO raises the signal level well above the noise level of the electronics circuit. Thus, the sensitivity of the coherent receiver is limited by the shot noise of the LO signal. Furthermore, because of the spatial mixing process, the coherent receiver is sensitive to signal and background noise only that falls within the same spatial temporal mode of the LO. This allows coherent detection optical receiver to operate in very strong background noise without significant degradation in the performance.

(b) Modulation format: The type of modulation format used for FSO link affects the sensitivity of the receiver. The coherent receiver, in particular heterodyne type, can be used for all kinds of modulation schemes. On the other hand, the homodyne receivers can be used only for intensity and phase-modulation schemes. The direct detection receiver can be used to detect only intensity-modulated signals, e.g., OOK or PPM. For long distance FSO link, a major concern is the higher receiver sensitivity at comparatively lower data rates. It is easily achievable with PPM scheme as it provides high peak-to-average power ratio. However, the PPM scheme is not bandwidth efficient. Therefore, except when the transmitter is peak power limited or when the system is modulation bandwidth limited, most of the FSO communication systems are based on PPM scheme [11].

(c) Photodetector and background noise: There are various types of noise sources which contribute noise in the FSO receiver. These noise sources include background noise, detector dark current noise, signal shot noise, and thermal noise. The contributions of these noise sources in the final output depend upon the optical design, receiver configuration, data bandwidth, and type of FSO link. A brief description on these noise sources is given below.

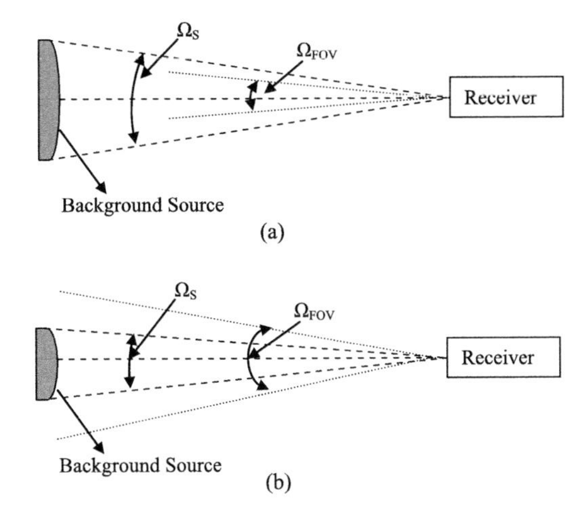

Fig. 3.5 Geometry of (**a**) extended source when $\Omega_{FOV} < \Omega_S$ and (**b**) stellar or point source when $\Omega_{FOV} > \Omega_S$

Background noise: The noise that is detected by the photodetector due to the surrounding environment is called background noise. The main sources of background noise are (a) diffused extended background noise from the atmosphere, (b) background noise from the Sun and other stellar (point) objects, and (c) scattered light collected by the receiver. Figure 3.5 shows the geometry of point and extended sources relative to the receiver. The background power collected by the receiver due to diffused, stellar, and scattered noise is given by

$$P_B = \begin{cases} H_B \Omega_{FOV} L_R A_R \Delta\lambda_{filter} & \text{(Diffused or extended source)} \\ N_B L_R A_R \Delta\lambda_{filter} & \text{(Stellar or point source)} \\ \gamma I_\lambda \Omega_{FOV} L_R A_R \Delta\lambda_{filter} & \text{(Scattered noise source)} \end{cases} . \tag{3.5}$$

In the above equation, H_B and N_B are the background radiance and irradiance energy densities of large extended angular sources and point sources, respectively. The term H_B is expressed in units of W/m^2/sr/Å and N_B in terms of W/m^2/Å. The parameters Ω_{FOV} is the solid angle of receiver field of view (FOV), L_R the transmission loss of the receiver optics, A_R the effective area of the receiver, and $\Delta\lambda_{filter}$ the bandwidth of optical BPF in the receiver. In case of scattered noise source, γ represents the atmospheric attenuation coefficient and I_λ the exo-atmospheric [2] (region of space outside the earth's atmosphere) solar constant (0.074 W/cm^{2}µm). A strong background source near the receiver FOV can lead to significant scattering. For an optical receiver design with optics under direct exposure to sunlight, the major background noise source is due to scattering. The contribution of celestial bodies such as the Sun, the moon, the star, etc., are assumed to be almost negligible except in deep space optical communication. With background noise power P_{BG}, the background noise current variance in an APD is given as

$$\sigma_{BG}^2 = 2qR_0 P_{BG} \mathcal{M}^2 F_B \quad (\text{A}^2). \tag{3.6}$$

In the above equation, B is the signal bandwidth. The parameter F is the excess noise factor which depends on the detector material and several other parameters. The commonly used expression for determining F as a function of APD gain, $\mathcal{M}$, is given below

$$F(\mathcal{M}) = \mathcal{M}^x, \tag{3.7}$$

where the value of x lies between $0 \leq x \leq 1.0$ and is dependent upon the material. The value of x is 0.3 for silicon, 0.7 for InGaAs, and 1.0 for Germanium photodetector. For PIN photodetector, the values of $\mathcal{M}$ and F are unity. The other parameters used in the equation are defined earlier.

Dark current noise: When there is no optical power incident on the photodetector, a small reverse leakage current still flows through the device. This is called dark current and contributes to total system noise. The detector dark current noise variance in an APD receiver is given by

$$\sigma_d^2 = 2q\mathcal{M}^2 F I_{db} B + 2q I_{ds} B \quad \left(\mathrm{A}^2\right). \tag{3.8}$$

In the above equation, I_{db} and I_{ds} are the bulk and surface dark currents, respectively. The dark current noise is dependent upon the operating temperature of the photodetector and its physical size. The dark currents can be reduced by cooling the detector or by reducing the physical size of the detector. It may be mentioned that avalanche multiplication is a bulk effect, and surface dark current is not affected by the avalanche gain, and therefore it can be neglected. In PIN photodetector, the dark current noise variance is given by

$$\sigma_d^2 = 2q I_d B \quad \left(\mathrm{A}^2\right). \tag{3.9}$$

The typical dark current values for some most commonly used detectors are given in Table 3.2

Signal shot noise: The number of photons emitted by an optical source at a given time is not constant. Therefore, the detection of photon by a photodetector is a discrete process (since the creations of electron-hole pairs result from the absorption of a photons) and is governed by the statistics of the photon arrival. The statistics of photons arriving at the detector follows the discrete probability distribution which is independent of the number of photons previously detected and is given by Poisson

Table 3.2 Typical values of dark current for various materials

Material	Dark current (nA)	
	PIN	APD
Silicon	1–10	0.1–1
Germanium	50–500	50–500
InGaAs	0.5–2.0	10–50

distribution. The fluctuations of the photons at the detector result in quantum or signal shot noise. The variance of signal shot noise current in an APD is expressed in terms of average signal photocurrent I_p as

$$\sigma_s^2 = 2q\mathcal{M}^2 F I_P B \quad \left(\mathrm{A}^2\right), \tag{3.10}$$

where $I_p = R_0 P_R$.

Thermal noise: This is the spontaneous fluctuations due to thermal interaction of electrons in any receiver circuit that consists of resistors at specific temperature. The thermal noise variance in the receiver due to resistor R_L can be expressed as

$$\sigma_{Th}^2 = \frac{4K_B T B}{R_L} \quad \left(\mathrm{A}^2\right), \tag{3.11}$$

where K_B is Boltzmann's constant, T the absolute temperature, and R_L the equivalent resistance of the circuit.

3.2.1 *Types of Detectors*

Based on the applications, the characteristics of the detector used are different and they can be classified into two types, i.e., communication and beacon detectors. The detectors used in FSO communication system for communication and beacon purposes are listed in Table 3.3.

(i) **Communication Detectors:** In an FSO communication system, commonly used detectors are PIN and APD. The APD provides moderate front end gain (50 to 200), low noise, and good quantum efficiency, depending upon the operating wavelength. The Si-APD has low noise, but suffers from low quantum efficiency above 1000 nm. However, InGaAs- and InGaAsP-based

Table 3.3 Communication and beacon detectors in FSO link

Application	Detector type	Material used
Communication	APD	Silicon, InGaAs, InGaAsP
	PIN	Silicon, InGaAs, InGaAsP
	CCD	Silicon
	PMT	Solid-state silicon photo-cathode
Beacon (acquisition and tracking)	CCD	Silicon
	CID	Silicon
	QAPD	Silicon
	QPIN	Silicon, InGaAs

APDs provide good quantum efficiency beyond 1000 nm. They are mostly used in the 1300 and 1550 nm operating wavelength range. However, the noise performance of these devices is very poor and as a result, Si devices normally outperform the longer wavelength devices even with low quantum efficiency. Besides APD, the PIN photodetector is also most widely used for direct detection communication. The gain of PIN photodetector is unity, but it provides good quantum efficiency. It is preferred for coherent applications where the required gain is provided by the LO. PIN photodetector can be fabricated on Si, InGaAs, and InGaAsP. At lower wavelengths, InGaAs and InGaAsP PIN photodetectors are used as they provide high quantum efficiency. Another type of detector used in early days was photomultiplier tubes (PMTs). These devices provide very high front-end gain (in the order of 10^5) and very low noise. However, due to heavy weight and large size, they are not suitable for the space-based applications. Another type of detector are charge-coupled devices (CCDs) which are good for low data rate as data rates depends upon the read-out speed of the CCD. Due to this reason, they are best suited as spatial acquisition detectors.

(ii) **Beacon Detectors:** Besides communication, ATP is also very essential for establishing an FSO communication link. Various types of spatial acquisition and tracking detectors are used at the receiver to detect the beacon signal that is searching over a large region of uncertainty. As discussed earlier, CCDs or charge injection devices (CIDs) configured as large array of detectors are used to perform both spatial acquisition and tracking functions for FSO communication system. The CCD-based array provides wide FOV and requires only one steering cycle. The FOV of beacon detector has to be chosen very carefully as large FOV will lead to large amount of background noise, and small FOV will result in long search patterns/time. Another type of detector used for detecting beacon signal is position-sensitive detector, e.g., quadrant APD and PIN detectors. The quadrant photodetectors are 2×2 array of individual photodetectors, separated by a small gap, fabricated on a single chip. This small spacing between the quadrants is also called dead zone. Ideally, the dead zone should be infinitely small, but in practice it is typically in the order of 50 to 100 μm. For large FOV, the spot size incident on the detector's dead zone could be lost. Therefore, instead of a finite dead zone between the quadrants, a transition across the boundaries of the quadrant is employed to avoid any abrupt change. These devices suffer from lot of crosstalk which degrade the performance. Quadrant APDs are generally used if receiver sensitivity requirement is high. Figure 3.6 shows the standard dead zone and sharing transition in quadrant APD.

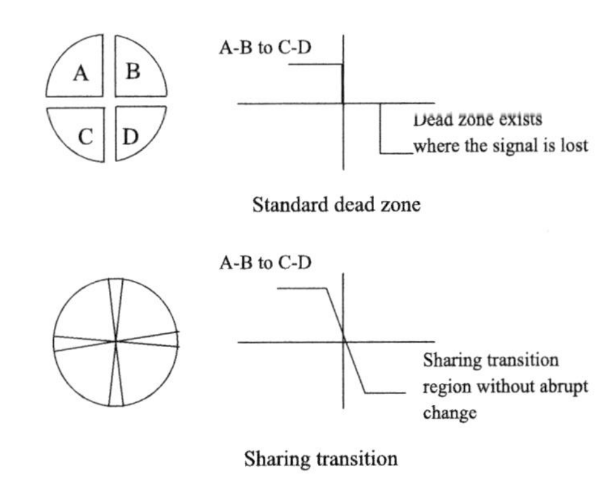

Fig. 3.6 Quadrant APD showing standard dead zone and shared transition

Table 3.4 Comparison between coherent and noncoherent receiver configurations

Parameters	Coherent	Noncoherent
Modulation parameters	Amplitude and phase	Intensity
Detection techniques	Heterodyne or homodyne detection	Direct detection (DD)
Adaptive control	Necessary	Not necessary

3.2.2 Receiver Configuration

The performance of any communication system for a given link distance and data rate depends upon receiver configuration as well as modulation scheme used. FSO communication uses either noncoherent (direct detection) or coherent (homodyne or heterodyne) receiver detection techniques. For intensity modulation, a simple receiver configuration, i.e., noncoherent detection, can be used. However, for digital modulation schemes like phase-shift keying (PSK) or frequency-shift keying (FSK), coherent receiver configuration is used which are more complex than noncoherent receiver configuration. Coherent receivers have advantage over noncoherent receivers in regard to their increased sensitivity level, data rates, and transmission distance. But due to their increased complexity in receiver design (as they have to restore the in-phase and quadrature components of the received optical fields/state of polarization of the signal) and difficult implementation, this configuration is not widely used in FSO communication system. Also, sometimes due to poor mixing of incoming signal and LO signal, the level of improvement in coherent receivers decreases significantly. Although coherent system is not a preferred choice for FSO communication, however, when dealing with very high data rates (100 Mbps or greater) or for power limited systems, these systems are given preference over direct detection system. Comparison between coherent and noncoherent receiver configurations is given in Table 3.4.

In this section, general expressions for output SNR in terms of required optical power, P_R in the presence of background noise for different receiver configurations, i.e., noncoherent detection (direct detection) or coherent detection (homodyne or heterodyne detection) are studied. Various receiver configurations considered are

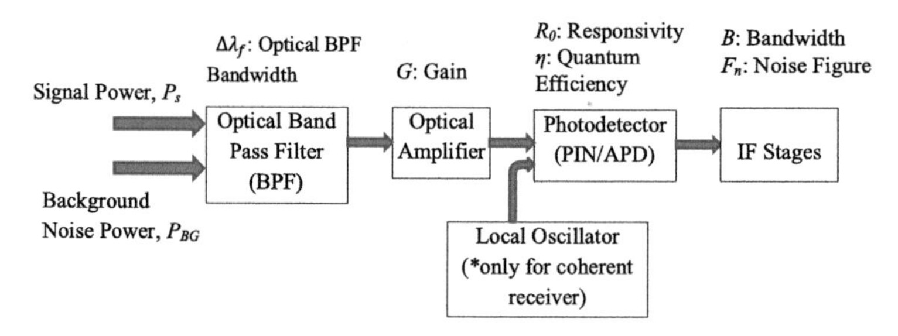

Fig. 3.7 A general optical communication receiver valid for all configurations

(i) coherent PSK homodyne receiver, (ii) coherent FSK heterodyne receiver, (iii) direct detection (PIN + Optical Amplifier) for OOK, (iv) direct detection APD for OOK, and (v) direct detection (APD) for $\mathbb{M}$-PPM. A general schematic diagram valid for all the abovementioned receiver configurations is shown in Fig. 3.7.

3.2.2.1 Coherent PSK Homodyne Receiver

The homodyne PSK receiver has the highest receiver sensitivity. In case of coherent receivers, photodetector in the receiver configuration act as a mixer for incoming optical signals and background noise (i.e., within the range of BPF). Therefore, it produces signal-background, background-background beat noise components along with the desired signal component. In addition to above noise components, the presence of LO will produce LO-background beat noise component too. All these noise components will collectively degrade the performance of FSO system. For homodyne receiver configuration, the frequency of LO is the same as that of incoming signal frequency.

For SNR analysis of PSK homodyne receiver, optical amplifier and IF stage blocks in Fig. 3.7 will not be used. The photocurrent is given as

$$i_p \propto |E_R(t) + E_{BG}(t) + E_L(t)|^2, \tag{3.12}$$

where E_s, E_{BG} and E_L are the electric field strengths due to signal, background, and LO, respectively. Following [7], various signal and noise components under the condition that optical bandwidth $\Delta\lambda_f$ is much greater than electrical bandwidth, $B(= 1/T_b$, where T_b is the bit duration), are given as

$$\text{Signal photocurrent}: i_p(t) = 2R_0\sqrt{P_R P_L}\cos\phi(t), \tag{3.13}$$

$$\text{DC current}: I_{dc} = R_0\left[P_R + P_L + m_t S_n \Delta\lambda_f\right], \tag{3.14}$$

$$\text{RMS noise current}: \overline{i_{bg}^2} = 2qI_{BG}B + R_0^2 S_n\left[2P_R + 2P_L + m_t S_n \Delta\lambda_f\right]2B + \frac{4K_B T B F_n}{R_L}, \tag{3.15}$$

where $\phi(t) = \phi_L - \phi_R$. The parameters P_R and ϕ_R are the power and phase of the received signal, respectively. Similarly, P_L and ϕ_L give the power and phase of LO signal, respectively. Other parameters, viz., R_0 is the responsivity, m_t are the number of background modes as seen by receiver FOV, S_n the background noise power spectral density per spatial mode, B the electrical bandwidth, $\Delta\lambda_f$ optical filter bandwidth, q electronic charge, I_{BG} background noise current, K_B Boltzmann's constant ($=1.38 \times 10^{-23}$ J/K), T absolute temperature in Kelvin, F_n noise figure of IF stages, and R_L photodetector load resistance. In Eq. (3.14), the first two terms arise due to received and LO signals, respectively. The third term represents DC component of the current arising due to background noise. In Eq. (3.15), the first term represents the shot noise. The second, third, and fourth terms represent the contribution due to signal-background, LO-background, and background-background beat noise, respectively. The last term in the equation represents the thermal noise contribution of the photodetector load resistance and the following stages.

The output SNR for binary PSK signaling scheme from Eqs. (3.13), (3.14), and (3.15) is given by

$$\frac{S}{N} = \frac{\overline{i_p^2(t)}}{\overline{i_{bg}^2}}. \tag{3.16}$$

After further simplification, the above equation becomes [5]:

$$\frac{S}{N} = \frac{2P_R}{h\nu BF_h}, \tag{3.17}$$

where

$$F_h = \frac{1}{\eta}\left[1 + \frac{P_R}{P_L} + \frac{S_n\Delta\lambda_f}{P_L}\right] + 2\frac{S_n}{h\nu}\left[1 + \frac{P_R}{2P_L} + \frac{S_n\Delta\lambda_f}{2P_L}\right] + \frac{K_BTF_n}{h\nu L_m}, \tag{3.18a}$$

and

$$L_m = \frac{1}{2}R^2P_LR_L. \tag{3.18b}$$

In the above equations, η is the quantum efficiency, and ν the frequency of received signal. In balanced homodyne receiver, signal-background and background-background noise and ASE-ASE components will cancel out. Therefore, Eq. (3.18a) reduces to

$$F_h = \frac{1}{\eta}\left[1 + \frac{P_R}{P_L} + \frac{S_n\Delta\lambda_f}{P_L}\right] + 2\frac{S_n}{h\nu} + \frac{K_BTF_n}{h\nu L_m}. \tag{3.19}$$

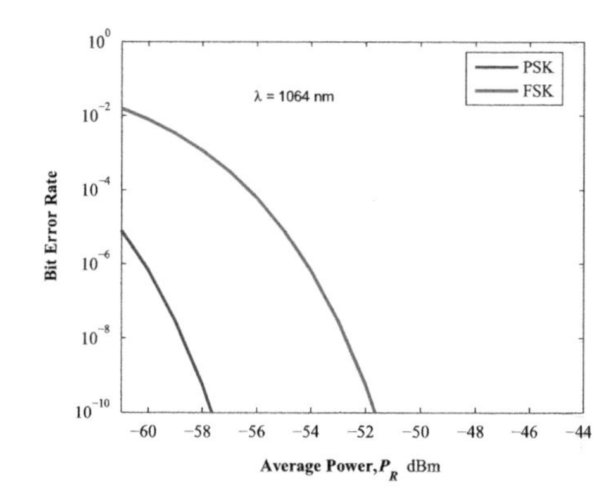

Fig. 3.8 Variation of P_e with average P_R for coherent receivers

The probability of error, P_e, is therefore given by:

$$P_e = \frac{1}{2} erfc\left(\frac{S}{N}\right)^{1/2}. \tag{3.20}$$

The variations of P_e with P_R are computed from the Eq. (3.20) for $\eta = 0.7, \Delta\lambda_f = 5$ nm, $S_n = 1.9 \times 10^{-20}$ W/Hz, $\lambda = 1064$ nm, $F_n = 2$ dB, $T = 300$ K, $R_L = 100\,\Omega$, B = 1 GHz (or data rate, $R_b = 1$ Gbps), and $P_L = 10$ dBm and are shown in Fig. 3.8.

3.2.2.2 Coherent FSK Heterodyne Receiver

For heterodyne receiver, intermediate frequency (IF) stage is used in Fig. 3.7. Here, the frequency of LO, ω_L, is offset from incoming signal frequency, ω_s, by a factor equal to intermediate frequency, ω_{IF}. Therefore, the received signal photocurrent at photodetector will be

$$\begin{aligned}\text{Signal photocurrent}: i_p(t) &= 2R_0\sqrt{P_R P_L}\cos\left[(\omega_{\mathrm{L}}(\mathrm{t}) - \omega_{\mathrm{s}}(\mathrm{t})) + \phi(t)\right] \\ &= 2R_0\sqrt{P_R P_L}\cos\left[\omega_{\mathrm{IF}}(\mathrm{t}) + \phi(t)\right].\end{aligned} \tag{3.21}$$

The analysis for coherent FSK heterodyne receiver is similar to that of coherent PSK homodyne receiver but the performance curve will shift toward the right by 6 dB. Figure 3.8 shows the probability error curve for orthogonal FSK signaling scheme.

3.2.2.3 Direct Detection (PIN + OA) Receiver for OOK

Direct detection (PIN + OA) receiver follows the same block diagram as shown in Fig. 3.7 except that IF stage and LO blocks will not be used here. Various signal and noise components under the same condition as in homodyne receiver, i.e., $\Delta\lambda_f \gg B$, are given as

$$\text{Signal photocurrent}: I_p = GR_0P_R, \tag{3.22}$$

$$\text{DC current}: I_{dc} = R_0\left[GP_R + GS_n\Delta\lambda_f + P_{sp}\Delta\lambda_f\right], \tag{3.23}$$

where G is the gain of optical amplifier, P_{sp} is the spontaneous noise power at the output of the amplifier and is given by

$$P_{sp} = n_{sp}(G-1)h\nu, \tag{3.24}$$

where n_{sp} is population inversion parameter of optical amplifier. Other parameters are same as mentioned above.

RMS noise current :

$$\begin{aligned}\sigma_1^2 \text{ for bit } '1' = {} & 2qI_{dc}B + R_0^2GS_n\left[2GP_R + GS_n\Delta\lambda_f\right]2B \\ & + R_0^2P_{sp}\left[2GP_R + P_{sp}\Delta\lambda_f + GS_n\Delta\lambda_f\right]2B \\ & + 4K_BTBF_n/R_L\end{aligned} \tag{3.25a}$$

and

$$\begin{aligned}\sigma_0^2 \text{ for bit } '0' = {} & qR_0\left(P_{sp}\Delta\lambda_f + GS_n\Delta\lambda_f\right)2B + R_0^2n_{sp}^2\Delta\lambda_f2B \\ & + R_0^2G^2S_n^2\Delta\lambda_f2B + 4K_BTBF_n/R_L.\end{aligned} \tag{3.25b}$$

For OOK signaling scheme, P_e for bit ‘1’ and ‘0’ are given by

$$P_{e1} = \frac{1}{2}erfc\left[\left(R_0GP_R - Th\right)/\sqrt{2}\sigma_1\right] \tag{3.26}$$

and

$$P_{e0} = \frac{1}{2}erfc\left[Th/\sqrt{2}\sigma_0\right], \tag{3.27}$$

where Th is the threshold level. If an optimum threshold level which equalizes P_{e1} and P_{e0} is used, then P_e from Eqs. (3.26) and (3.27) will be

$$P_e = P_{e1} = P_{e0} = \frac{1}{2}erfc\left(S/2N\right)^{1/2}, \tag{3.28}$$

where

$$\frac{S}{N} = \frac{R_0^2G^2P_R^2}{\left(\sigma_1 + \sigma_0\right)^2}. \tag{3.29}$$

The SNR for this receiver configuration after simplification is given as [5]

$$\frac{S}{N} = \frac{P_R}{2h\nu B\left(\sqrt{F_{p1}} + \sqrt{F_{p0}}\right)^2}, \tag{3.30}$$

where

$$F_{p1} = \frac{1}{\eta G}\left[1+\frac{S_n\Delta\lambda_f}{P_R} + \frac{n'_{sp}h\nu\Delta\lambda_f}{P_R}\right] + 2\left[n'_{sp} + \frac{S_n}{h\nu}\right]\cdot\left[1 + \frac{S_n\Delta\lambda_f}{2P_R} + \frac{n'_{sp}h\nu\Delta\lambda_f}{2P_R}\right] + \frac{K_BTF_n}{h\nu G^2L'_m} \tag{3.31}$$

and

$$F_{p0} = \frac{1}{\eta G}\left[\frac{S_n\Delta\lambda_f}{P_R} + \frac{n'_{sp}h\nu\Delta\lambda_f}{P_R}\right] + 2\left[n'_{sp} + \frac{S_n}{h\nu}\right]\cdot\left[\frac{S_n\Delta\lambda_f}{2P_R} + \frac{n'_{sp}h\nu b}{2P_R}\right] + \frac{K_BTF_n}{h\nu G^2L'_m}. \tag{3.32}$$

The other parameters are

$$n'_{sp} = \left(1 - \frac{1}{G}\right)n_{sp} \tag{3.33}$$

and

$$L'_m = \frac{1}{2}R_0^2P_RR_L. \tag{3.34}$$

All other parameters in Eqs. (3.30), (3.31), (3.32), (3.33), and (3.34) are defined earlier and are same as in PSK homodyne receiver. The variation of P_e with average P_R for (PIN + OA) direct detection receiver at $\lambda = 1064$ nm is shown in Fig. 3.9.

3.2.2.4 Direct Detection (APD) Receiver for OOK

This receiver configuration follows the same block diagram as shown in Fig. 3.7 except that IF stage and LO blocks will not be there. The detector will be APD instead of PIN. In this case, various signal and noise components under the condition $\Delta\lambda_f \gg B$ are given by

$$\text{Signal photocurrent}: I_p = R_0\mathcal{M}P_R, \tag{3.35}$$

$$\text{DC current}: I_{dc} = \mathcal{M}^2FR_0\left[P_R + S_n\Delta\lambda_f\right], \tag{3.36}$$

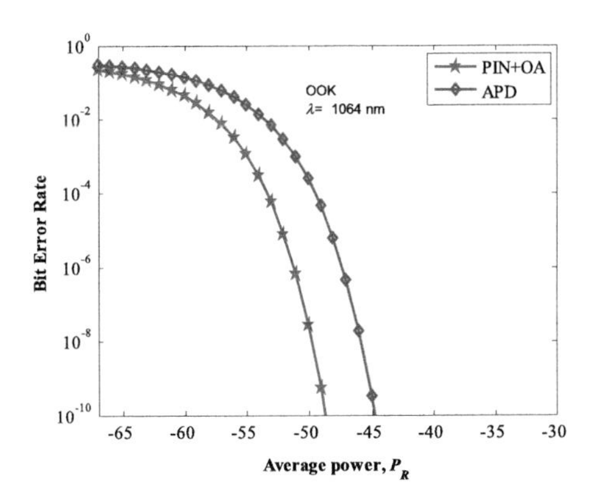

Fig. 3.9 Variation of P_e with average P_R for direct detection receivers

RMS noise current :

$$\sigma_1^2 \text{ for bit '1'} = 2qI_{dc}B + \mathcal{M}^2 R_0^2 S_n \left[2P_R + S_n \Delta\lambda_f\right] 2B + 4K_B TBF_n/R_L \tag{3.37}$$

and

$$\sigma_0^2 \text{ for bit '0'} = q\mathcal{M}^2 F R_0 S_n \Delta\lambda_f 2B + \mathcal{M}^2 R_0^2 S_n^2 \Delta\lambda_f 2B + 4K_B TBF_n/R_L. \tag{3.38}$$

where $\mathcal{M}$ and F are multiplication gain and excess noise figure of APD as defined earlier. Following the same approach as in direct detection (PIN + OA) receiver for OOK, SNR can be analyzed by using Eq. 3.30 with F_{p1} and F_{p0} replaced by F_{a1} and F_{a0}, respectively. F_{a1} and F_{a0} are given by [5]

$$F_{a1} = \frac{F}{\eta}\left[1 + \frac{S_n \Delta\lambda_f}{P_R}\right] + 2\frac{S_n}{h\nu}\left[1 + \frac{S_n \Delta\lambda_f}{2P_R}\right] + \frac{K_B T F_n}{h\nu \mathcal{M}_{opt}^2 L_m'} \tag{3.39}$$

and

$$F_{a0} = \frac{F}{\eta}\left[\frac{S_n \Delta\lambda_f}{P_R}\right] + 2\frac{S_n}{h\nu}\left[\frac{S_n \Delta\lambda_f}{2P_R}\right] + \frac{K_B T F_n}{h\nu \mathcal{M}_{opt}^2 L_m'}, \tag{3.40}$$

where

$$\mathcal{M}_{opt}^{x+2} = \frac{4K_B TF_n/R_L}{xqR_0\left(P_R + S_n \Delta\lambda_f\right)}. \tag{3.41}$$

$\mathcal{M}_{opt}$ is the optimum value of multiplication gain, $\mathcal{M}$ which maximizes SNR. All other parameters are same as defined earlier. The variation of P_e with average P_R for (APD + OOK) direct detection receiver for $x = 0.5$ is shown in Fig. 3.9.

3.2.2.5 Direct Detection (APD) for $\mathbb{M}$-PPM

As discussed earlier in Chap. 1, in $\mathbb{M}$-PPM scheme, each word contains n bits of information, i.e., $\mathbb{M} = 2^n$. Therefore, a complete data word is divided into $\mathbb{M}$ slots of duration T_s seconds, and information is placed in any one of these slots. The bandwidth occupied by $\mathbb{M}$-PPM is given by

$$B_{PPM} = \frac{1}{T_s}, \tag{3.42}$$

where

$$T_s = \frac{T_b \log_2 \mathbb{M}}{\mathbb{M}}. \tag{3.43}$$

Hence, it is seen that bandwidth occupied by PPM is more than bandwidth occupied by OOK $(= \frac{1}{T_b})$. Using the same approach as in OOK, the probability of error P_e for bits '1' and '0' in 2-PPM scheme are given by

$$P_e = P_{e1} = P_{e0} = \frac{1}{2} erfc \left[\frac{R_0^2 \mathcal{M}^2 P_R^2}{2\left(\sigma_1^2 + \sigma_0^2\right)} \right]^{1/2}. \tag{3.44}$$

This equation can be written in terms of SNR as

$$\frac{S}{N} = \frac{R_0^2 \mathcal{M}^2 P_R^2}{\left(\sigma_1^2 + \sigma_0^2\right)}. \tag{3.45}$$

Substituting σ_1^2 and σ_0^2 from Eqs. (3.37) and (3.38), respectively, and after further simplification and replacing B with B_P, the above equation can be written as

$$\frac{S}{N} = \frac{P_R}{2h\nu B_{PPM}\left(F_{a1} + F_{a0}\right)}. \tag{3.46}$$

In $\mathbb{M}$-PPM, decision regarding the presence of a pulse in a time slot is made on the basis of $\mathbb{M} - 1$ comparisons. The probability of wrongly decoding a word, P_{ew}, is given by

$$P_{ew} = 1 - \int_{-\infty}^{\infty} \left[\int_{-\infty}^{(R_0 \mathcal{M} P_R + I_1)} p\left(I_0\right) dI_0 \right]^{m-1} p\left(I_1\right) dI_1, \tag{3.47}$$

where $p(I_1)$ and $p(I_0)$ are the probability density functions of I_1 and I_0, respectively. These are considered to be Gaussian distributed with variance σ_1^2 and σ_0^2, respectively. After simplification, Eq. (3.47) reduces to

$$P_{ew} = 1 - \frac{1}{\sqrt{2\pi\sigma_1^2}} \int_{-\infty}^{\infty} P\left[1 - \frac{1}{2} erfc \frac{R_0 \mathcal{M} P_R + I_1}{\sqrt{2}\sigma_0}\right]^{\mathbb{M}-1} \cdot \exp\left[-\frac{I_1^2}{2\sigma_1^2}\right] dI_1. \tag{3.48}$$

This P_{ew} can be used to obtain the upper bound on the probability of error, P_e. Relationship between P_{ew} and P_e is given as

$$P_e \leq \frac{\mathbb{M}/2}{\mathbb{M}-1} P_{ew}. \tag{3.49}$$

If the second term inside the square bracket in Eq. (3.48) is very small ($\ll 1$), then Eq. (3.48) can be approximated as

$$P_{ew} \approx \frac{\mathbb{M}-1}{2} erfc \left[\frac{R_0^2 \mathcal{M}^2 P_R^2}{2\left(\sigma_1^2 + \sigma_0^2\right)}\right]^{1/2}. \tag{3.50}$$

3.3 Optical Post and Preamplifiers

Both post-amplifier and preamplifier, in addition of amplifying the input signal, will add to it a signal due to spontaneous emission of light. A portion of this spontaneous emitted light is in the same direction as the signal and gets amplified along with the main signal. This added light, called the amplified spontaneous emission (ASE) noise, is spread over a wide frequency range. The spontaneous noise power at the output of the amplifier is given by

$$P_{sp} = (G-1)\, n_{sp} h\nu B_o, \tag{3.51}$$

where n_{sp} is spontaneous emission factor (or sometimes called population inversion factor), G the amplifier gain, h the Planck's constant, ν the incoming frequency, and B_o the optical filter bandwidth. The output power given by Eq. (3.51) is for each polarization mode. Since we have two fundamental polarization modes, the total noise power at the output of the amplifier will be $2P_{sp}$. The value of n_{sp} depends upon the population inversion and is given by

$$n_{sp} = \frac{N_2}{N_2 - N_1}, \tag{3.52}$$

where N_1 and N_2 are the atomic population in ground and excited states, respectively.

The amplified output signal plus ASE noise when detected by a photodetector consists of desired signal component along with the thermal, shot, and beat noise components. Among shot noise are signal shot noise, ASE shot noise, and background shot noise. The beat noise components consists of signal-background beat noise, background-background beat noise, ASE-ASE beat noise, and signal-ASE beat noise. The variances of all the noise components, namely, thermal noise, shot noise, signal-ASE beat noise, ASE-ASE beat noise, and amplified background noise, at the receiver output are given by [10, 12]:

$$\sigma_{th}^2 = \frac{4K_B T F_n B_e}{R_L}, \tag{3.53}$$

$$\sigma_{sig-shot}^2 = 2qR_0 G P_R B_e, \tag{3.54}$$

$$\sigma_{ASE-shot}^2 = 2qI_{sp}B_e = 2qR_0 P_{sp} B_e = 2qR_0 \left(G-1\right) n_{sp} h\nu B_e B_o \tag{3.55}$$

and

$$\sigma_{sig-BG}^2 = 2qI_{BG}B_e = 2qR_0 G P_{BG} B_e. \tag{3.56}$$

Combining Eqs. (3.54), (3.55), and (3.56), total variance due to shot noise is given by

$$\sigma_{shot}^2 = \sigma_{sig-shot}^2 + \sigma_{ASE-shot}^2 + \sigma_{sig-BG}^2, \tag{3.57}$$

$$\sigma_{shot}^2 = 2qR_0 \left(GP_R + \left(G-1\right) n_{sp} h\nu B_o + GP_{BG}\right) B_e. \tag{3.58}$$

Variance of various beat noise components are given as

$$\sigma_{sig-ASE-beat}^2 = 4R_0^2 G P_R n_{sp} h\nu \left(G-1\right) B_e, \tag{3.59}$$

$$\sigma_{sig-BG-beat}^2 = 4R_0^2 G^2 P_R \left(n_{sp} h\nu \left(G-1\right)\right) B_e, \tag{3.60}$$

$$\sigma_{ASE-ASE-beat}^2 = 2R_0^2 \left[n_{sp} h\nu \left(G-1\right)\right]^2 \left(2B_o - B_e\right) B_e, \tag{3.61}$$

$$\sigma_{BG-BG-beat}^2 = R_0^2 N_{BG}^2 G^2 \left(2B_o - B_e\right) B_e, \tag{3.62}$$

$$\sigma_{ASE-BG-beat}^2 \approx 2R_0^2 G N_{BG} \left(n_{sp} h\nu \left(G-1\right)\right) \left(2B_o - B_e\right) B_e. \tag{3.63}$$

In the above equations, N_{BG} is the background power spectral densities (W/Hz) at the input of optical preamplifier. Therefore, total variance is given as

$$\sigma_{total}^2 = \sigma_{shot}^2 + \sigma_{sig-ASE-beat}^2 + \sigma_{sig-BG-beat}^2 + \sigma_{ASE-ASE-beat}^2 + \sigma_{BG-BG-beat}^2 + \sigma_{ASE-BG-beat}^2 + \sigma_{th}^2. \tag{3.64}$$

Since the amplifier gain is reasonably large, the contribution due to shot noise and thermal noise are negligible as compared to signal-ASE and ASE-ASE beat noise. The ASE-ASE noise can be made very small by reducing the optical bandwidth, B_o. Therefore, the dominant noise component is usually signal-ASE beat noise. In that case, the SNR at the amplifier output is given by the following equation:

$$SNR_o = \frac{(R_0 G P_R)^2}{\sigma^2_{sig-ASE-beat}} = \frac{(R_0 G P_R)^2}{4R^2 G P n_{sp} h\nu \,(G-1)\, B}. \tag{3.65}$$

In the above equation, B is the electrical bandwidth which is same as B_e.

3.4 Link Design Trade-Off

In the design of FSO link for a given requirements, some trade-offs have to be made among various design parameters. These are discussed below.

3.4.1 Operating Wavelength

The choice of operating wavelength depends upon many factors that includes:

(i) Availability of laser: While choosing the laser for any system, we need to consider peak-to-average power ratio, available peak power, electrical-to-optical conversion efficiency, and overall power consumption. Therefore, the trade-off between available laser technologies which strongly depend upon operating wavelength should be made to identify the appropriate choice.

(ii) Gain vs. beamwidth: In general, the gain of optical transmit or receive antenna is given as $G \approx (\pi D_R/\lambda)^2$. Thus gain is inversely proportional to the operating wavelength, and hence it is desirable to work at lower operating wavelengths to get more gain. However, beamwidth of the system is proportional to (λ/D). This implies that at lower operating wavelength, beamwidth will be narrower leading to increase in pointing errors. Therefore, a trade-off between higher gain and reducing signal fades due to pointing error has to be considered.

(iii) Atmospheric absorption and scattering: The atmospheric absorption and scattering depend upon the choice of operating wavelength. When a light beam travels through the atmosphere, it may be absorbed or scattered by the constituent particles of the atmosphere. Only the wavelengths outside the main absorption band can be used for optical communication. The region of maximum absorption is called forbidden band, and the region used for optical communication is called atmospheric transmission band. The transmission window for FSO communication system is in the visible and near-infrared region that stretches roughly from 750 to 1600 nm. However, certain wavelengths in the near-infrared region suffer from strong atmospheric absorption due to the presence of the water particles (moisture). The contribution of gas absorption due to oxygen and carbon dioxide to overall absorption coefficient can be neglected as the gas-specific absorption coefficient is very small as compared to water absorption. However, at longer infrared wavelength range (>2000 nm), gas absorption can dominate the absorption properties of the atmosphere. There are several transmission windows that are nearly transparent between 750 and 1600 nm. These are 850, 1060, 1250, and 1550 nm. The 850 and 1064 nm are characterized by low attenuation windows with reliable and inexpensive transmitter and receiver components. However, at 1064 nm, Nd/YAG laser is preferred, and because of this FSO systems at this wavelength are bulky. The 1250 nm window is very rarely used as it shows a drastic increase in absorption at 1290 nm leaving a very small workable band. The 1550 nm is very well suited for high-quality and low-attenuation transmission. The components in this range are easily available and reliable. The background noise contribution at the receiver output decreases with the increase in the operating wavelength. Consequently, 1550 nm wavelength is less vulnerable to background noise.

(iv) Detector sensitivity: The sensitivity of the PIN detector is determined by its detection efficiency. In case of APD, the detection sensitivity depends upon gain $\mathcal{M}$, quantum efficiency η, and excess noise factor F. A good APD detector is characterized by high gain, large bandwidth, high efficiency, and low excess noise factor. However, the availability of the detector is very much limited by the operating wavelength. For example, Si detectors can provide a very high gain bandwidth and low excess noise factor, but they have less detection sensitivity at 1500 nm.

3.4.2 *Aperture Diameter*

The power efficiency in FSO link depends on the transmitter and receiver aperture areas. In order to reduce the transmit power requirement, it is desirable to have larger receiver aperture size. However, the receiver aperture area cannot be increased indefinitely as it will enhance the background noise contribution and leads to increase in mass of the terminal. Further, the size of aperture area affects the pointing requirement. Since the beamwidth is inversely proportional to the transmitter aperture diameter, larger aperture size will require tighter pointing accuracy and higher sensitivity toward pointing loss. Therefore, an optimum choice of aperture diameter has to be made in order to increase the power efficiency in FSO system.

3.4.3 *Receiver Optical Bandwidth*

The FSO link performance can be improved by reducing the background noise with the help of BPFs. The filter bandwidth should be sufficient enough to pass the information signal without any distortion. It should not be very wide, otherwise

Table 3.5 Parameters for FSO link design

S.No.	Link budget	Parameters
1.	Transmitter parameters	Operating wavelength
		Transmit power
		Transmitter aperture area
		Transmitter optical efficiency
		Transmitter antenna gain
2.	Channel losses and noise	Pointing loss
		Atmospheric losses
		Scintillation-induced loss
		Atmospheric background noise
		Free-space loss
3.	Receiver parameters	Receiver aperture area
		Receiver sensitivity
		Receiver optics efficiency
		Receiver detector FOV
		Receiver antenna gain
		Bandwidth of narrow BPF
		Optical post and preamplifier emission noise, i.e., ASE noise

background noise contribution will increase. Therefore, a BPF of appropriate bandwidth is to be designed keeping in view the practical constraints.

Various parameters which are to be considered in the FSO link design are given in Table 3.5.

3.5 Summary

The performance of FSO communication system depends directly on the efficiency and sensitivity of optical components, i.e., optical transmitter, optical modulator, optical amplifier, and optical receiver used in link design. The fundamental characteristics of these optical components and associated noise sources are discussed in this chapter. Receivers with both direct detection (average power limited) and coherent detection (high power efficient) using various modulation format are presented. The minimum required average optical power, i.e., receiver sensitivity for achieving a given performance at a certain data rate, has been evaluated for PIN and APD receivers for PSK, FSK, and OOK modulation formats. Link design trade-offs has been discussed in order to provide a fair understanding of design for FSO communication system.

Bibliography

1. J.M. Budinger, M.J. Vanderaar, P.K. Wagner, S.B. Bibyk, Combinatorial pulse position modulation for power-efficient free-space laser communications, in Proceedings of SPIE (1993), pp. 214–225
2. A. Farrell, M. Furst, E. Hagley, T. Lucatorto et al., Surface and exo-atmospheric solar measurements. Technical report, Physical Measurement Laboratory
3. Z.Z. Ghassemlooy, A.R. Hayes, B. Wilson, Reducing the effects of intersymbol interference in diffuse DPIM optical wireless communications. Optoelectron. IEE Proc. **150**(5), 445–452 (2003)
4. H. Hemmati, *Near Earth Laser Communications* (CRC Press/Taylor & Francis Group, Boca Raton, 2009)
5. V.K. Jain, Effect of background noise in space optical communication systems. *Int. J. Electron. Commun. (AEÜ)* **47**(2), 98–107 (1993)
6. S.G. Lambert, W.L. Casey, *Laser Communication in Space* (Artech House, Boston, 1995)
7. W.R. Leeb, Degradation of signal-to-noise in optical free space data link due to background illumination. Appl. Opt. **28**, 3443–3449 (1989)
8. T. Ohtsuki, I. Sasase, S. Mori, Lower bounds on capacity and cutoff rate of differential overlapping pulse position modulation in optical direct-detection channel. IEICE Trans. Commun. **E77-B**, 1230–1237 (1994)
9. Practical uses and applications of electro-optic modulators. NEW FOCUS, Inc, Application Note 2. [Weblink: https://www.newport.com/n/practical-uses-and-applications-of-electro-optic-modulators]
10. R. Ramaswani, K.N. Sivarajan, *Optical Networks: A Practical Perspective* (Morgan Kaufmann, San Francisco, 2002)
11. D. Shiu, J.M. Kahn, Differential pulse-position modulation for power efficient optical communication. IEEE Trans. Commun. **47**(8), 1201–1210 (1999)
12. P.J. Winzer, A. Kalmar, W.R. Leeb, Role of amplifed spontaneous emission in optical free-space communication links with optical amplifcation-impact on isolation and data transmission and utilization for pointing, acquisition, and tracking. Proc. SPIE Free-Space Laser Commun. Technol. XI **3615**, 134 (1999)

Chapter 4
Acquisition, Tracking, and Pointing

4.1 Acquisition Link Configuration

Spatial acquisition of the onboard satellite terminal using a narrow laser beam is a very difficult task. Before the actual data transmission begins, the ground-based receiving terminal is first required to establish a LOS link to the satellite. It can be accomplished using a beacon laser signal of sufficient beam divergence that allows the transmitted power to search within the uncertainty area (in the order of mrad) of the receiver. The uncertainty area is typically larger than the beam divergence required for detection. This beacon signal is then acquired by the onboard satellite which is simultaneously searching for the beacon signal in its FOV. This is done with the help of focal pixel array (FPA) which has a FOV sufficiently wide to cover the full search FOV. Hence, the receiving terminal need not scan its own uncertainity area to acquire the beacon signal and thus helps in reducing the acquisition time. Once the signal is received, the controller logic on the satellite begins the process of narrowing its FOV until both systems have locked on to each other's signal. Controller logic then commands the optical beam steering element to keep the received signal bore sighted on the detector [1]. At this point, tracking loop is activated and onboard laser is then turned on for the downlink. The downlink beam is then picked up by the ground-based receiver to complete the link. Once the link is established, the transmission of data from the onboard laser can take place. In order to minimize the time required to acquire the target and improve system efficiency, the narrow laser beam should steer or point synchronously toward the receiver FOV. If the transmitter laser beam is larger than the receiver FOV, it will result in energy loss. If the laser beam is narrower than the receiver FOV, it increases the acquisition time. Therefore, in order to reduce the acquisition time, an efficient acquisition process is needed to rapidly search within the receiver FOV.

The detection of beacon and data signals is accomplished with the help of FPA which is designed to work in two modes. For detecting a beacon signal,

H. Kaushal et al., *Free Space Optical Communication*, Optical Networks,
DOI 10.1007/978-81-322-3691-7_4

Phase 1: Transmitter initiates the acquisition process by transmitting a beacon signal that scans over an uncertainty region and the location is acquired by the receiver

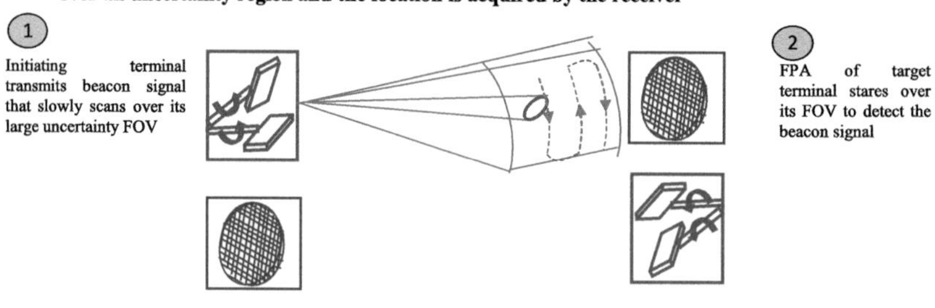

Phase 2: Receiver responds back and its location as well identity is acquired by the transmitter

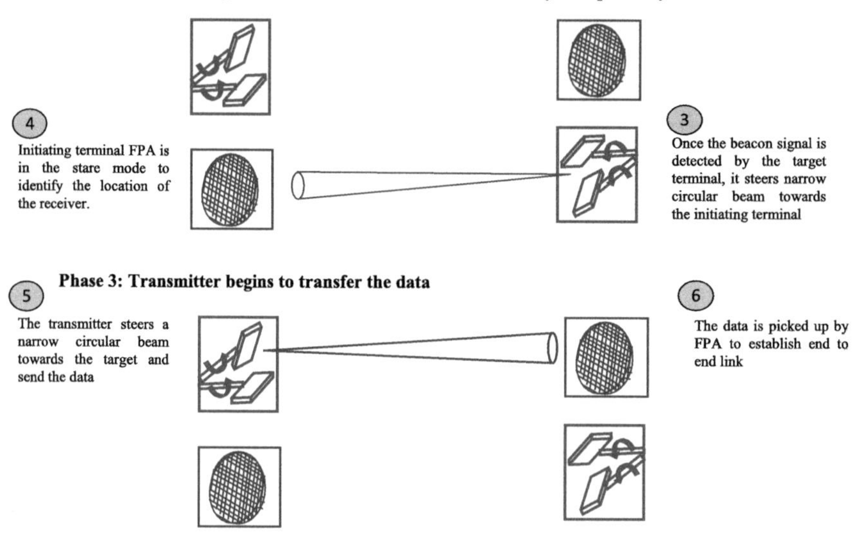

Fig. 4.1 Concept of acquisition link establishment between initiating and target parties

it operates in a “stare” mode where it detects the incoming beam and monitors every pixel in detector array continuously. For detecting data signal, it is switched electronically to “data receiving mode” where only those pixels are monitored where the incoming beacon signal has formed an image on FPA. The output of these pixels is preamplified and sent to data detection circuit. Since all other pixels are deactivated, it will enable FPA to serve as high-speed and low-noise receiver. The phase-wise description of link establishment and setting up a link between initiating ground-based station and onboard target parties is depicted in Fig. 4.1.

There is an important term used in ATP system, i.e., “point ahead angle” (PAA) [2]. This is one of the critical parameter in inter-satellite or ground-to-satellite laser communication that arises due to relative angular velocity of two terminals. The PAA angle can be achieved to a high degree of accuracy with the help of ephemerides data (the position of satellite according to orbit equation). The PAA

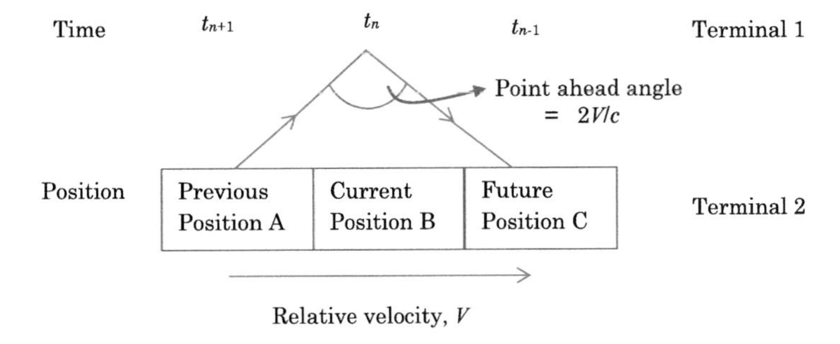

Fig. 4.2 Concept of point ahead angle

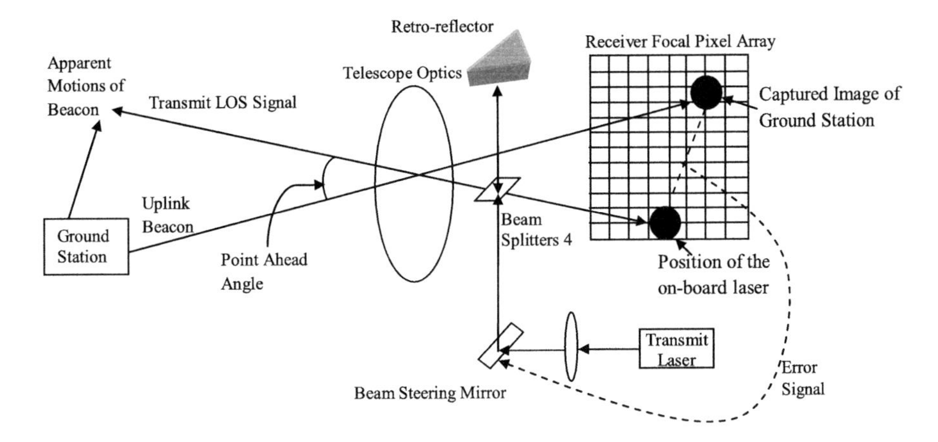

Fig. 4.3 Illustration of point ahead angle in FSO communication system

as shown in Fig. 4.2 will correct the relative LOS so that the transmitted signal effectively reaches the targeted terminal. Figure 4.3 illustrates the concept of PAA. In this figure, the signal from the laser strikes two-axis beam steering mirror. After reflecting off the beam steering mirror, it passes up to the dichroic beam splitter 4 that reflects out almost all the signal through the telescope. However, a small fraction of the signal passes through the dichroic beam splitter 4 and progresses upward to a retroreflector. The retroreflected signal is returned back and is allowed to focus an image on the FPA. This image represents the direction of the outgoing laser beam relative to the telescope axis. At a distance R from the ground station, the PAA is given by

$$\text{Point Ahead Angle (PAA)} = \underbrace{\frac{2R}{c}}_{\text{Round Trip Time}} \times \frac{\text{Projected velocity}}{R}, \tag{4.1}$$

where c is the velocity of light. It is seen from above that PAA is independent of distance between onboard satellite and ground station. In case of a LEO satellite moving at a speed of 7 km/hr, the PAA will be nearly 50 μrad.

For this purpose, a scanning system is required to rapidly scan the transmitted beam and receiver FOV. Various issues involved in the configuration of acquisition link are discussed in the following subsections.

4.1.1 Acquisition Uncertainty Area

Acquisition requires searching the uncertainty area to locate and establish a link between ground station and distant spaceborne satellite. A large uncertainty area and a narrow beam divergence can lead to a unreasonably long acquisition time. The initial uncertainty area in terms of solid angle depends upon different errors [3, 4]. These errors are a combination of attitude and ephemeris uncertainties in the satellite navigation system and are expressed as azimuth and elevation uncertainties. Various contributors to uncertainty area along with their typical values are given in Fig. 4.4. These individual error contributors are usually uncorrelated and random in nature. With these values, the uncertainty area in terms of solid angle is approximately 1 mrad.

Since the acquisition process is a statistical process, an appropriate mathematical model is required to determine the acquisition probability and mean acquisition time of a single scan. This will include the distribution function of satellite position and the time for scan.

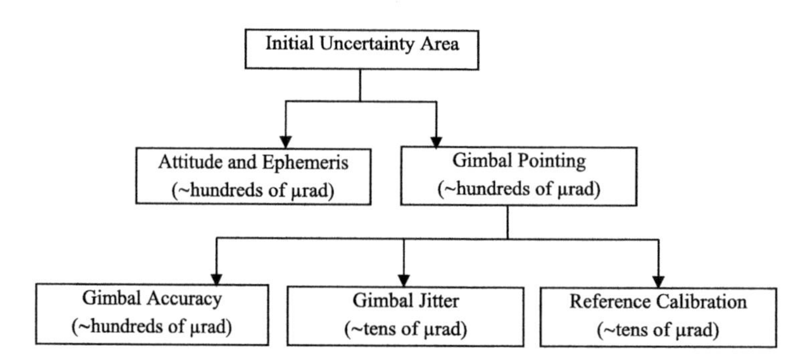

Fig. 4.4 Various contributors to the acquisition initial uncertainty area budget for ground-to-satellite FSO link

4.1.1.1 Probability Distribution Function of Satellite Position

The pointing error caused by satellite distribution can be described statistically by Gaussian distribution for both azimuth and elevation uncertainties [5]. The probability density function (pdf) of elevation pointing error angle which is normally distributed is given as

$$f(\theta_V) = \frac{1}{\sqrt{2\pi}\sigma_V}\exp\left[-\frac{(\theta_V - \mu_V)^2}{2\sigma_V^2}\right], \tag{4.2}$$

where θ_V is the elevation pointing error angle, μ_V the mean value, and σ_V the standard deviation. The pdf of azimuth pointing error angle is given by

$$f(\theta_H) = \frac{1}{\sqrt{2\pi}\sigma_H}\exp\left[-\frac{(\theta_H - \mu_H)^2}{2\sigma_H^2}\right], \tag{4.3}$$

where θ_H is the azimuth pointing error angle, μ_H the mean value, and σ_H the standard deviation. The radial pointing error angle (without bias) is the root sum square of the elevation θ_V and azimuth θ_H angles given by

$$\theta = \sqrt{\theta_H^2 + \theta_V^2}. \tag{4.4}$$

For simplicity, we assume the standard deviation of elevation and azimuth angles are equal so that

$$\sigma_\theta = \sigma_V = \sigma_H, \tag{4.5}$$

where σ_θ is the pointing error angle standard deviation. Further, assuming that the azimuth and elevation processes are zero-mean, independent, and identically distributed, the radial pointing error angle can be modeled as a Rician density distribution function given by [6]

$$f(\theta,\phi) = \frac{\theta}{\sigma_\theta^2}\exp\left(-\frac{\theta^2 + \phi^2}{2\sigma_\theta^2}\right) I_0\left(\frac{\theta\phi}{\sigma_\theta^2}\right), \tag{4.6}$$

where I_0 is the modified Bessel function of order zero and ϕ the bias error angle from the center. When ϕ is zero, the above equation leads to the well-known Rayleigh distribution function for pointing error angles and is given by [6]

$$f(\theta) = \frac{\theta}{\sigma_\theta^2}\exp\left(-\frac{\theta^2}{2\sigma_\theta^2}\right). \tag{4.7}$$

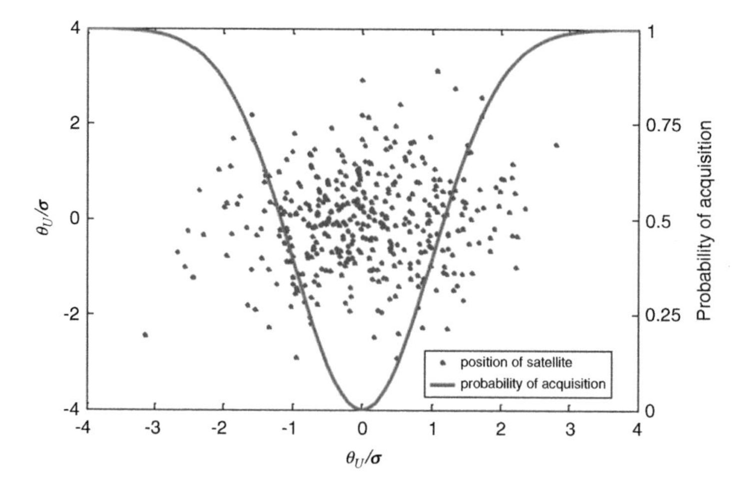

Fig. 4.5 Probability of acquisition as a function of the ratio of half-width of uncertainty area, θ_U, to the deviation of satellite position, σ [7]

This Rayleigh distributed function is used to evaluate the acquisition probability. The transmitter field of uncertainty (FOU) that is defined as the size of uncertainty area scanned for the target satellite usually depends on the deviation of target satellite's position, and it decides the probability of acquisition. In order to bring the probability of acquisition to an acceptable level and reduce the acquisition time, it is essential to determine the optimum size of FOU. The acquisition probability is given as

$$P_{acq} = \int_0^{\theta_U} f(\theta)\, d\theta = 1 - \exp\left(-\frac{\theta_{\mathrm{U}}^2}{2\sigma^2}\right), \tag{4.8}$$

where θ_U is the half-width of FOU angle. In Fig. 4.5, each scattered spot represents a possible location of satellite in transmitter satellite's FOU. The solid line represents the corresponding acquisition probability for various values of θ_U/σ. It is seen from this figure that in order to achieve a high degree of acquisition probability, the value of $\theta_U/\sigma = 3$. For this reason, 3σ level is generally adopted by acquisition system design engineers for a single scan.

4.1.2 Scanning Techniques

Scanning can be performed in various ways. Most popularly used scanning techniques are spiral and raster scan. Spiral scan can further be classified into continuous

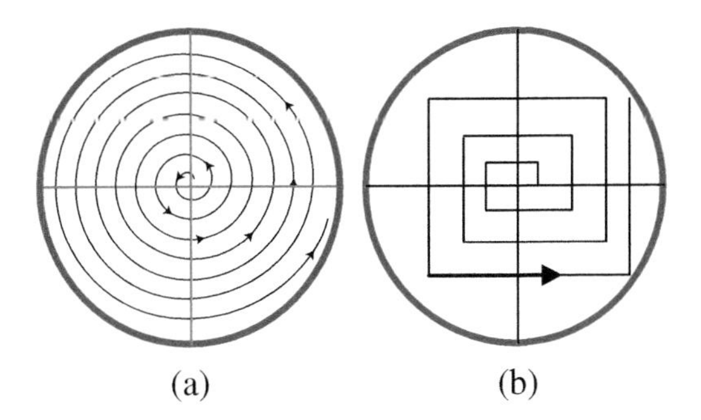

Fig. 4.6 Spiral scan pattern (**a**), continuous spiral scan, and (**b**) step spiral scan

spiral, step spiral, segmented spiral, and multiple spiral. These scanning techniques are discussed in details in this section.

(i) Continuous spiral scan: The spiral scan is the most efficient spatial acquisition technique as it is very easy to implement. The trace of spiral scan can be described in polar coordinates as

$$r_s = \frac{L_\theta}{2\pi}\theta_s, \tag{4.9}$$

where L_θ is the step length related to beacon beam divergence angle (θ_{div}) as $L_\theta = \theta_{div}(1 - F_o)$ where F_o is the overlap factor. It is clear from this equation that as the beam divergence increases, the step length also increases and hence the trace of spiral scan increases which reduces the acquisition time. Further, if beam divergence increases, then it requires large transmitted power and a large-size telescope. This will make the system costly and more complex. Figure 4.6 shows the continuous spiral scan pattern.

The scan time from the initial point (0, 0) to (r_s, θ_s) is given as

$$T(r_s) \approx \frac{\pi\theta_s^2}{L_\theta^2}T_{dwell}, \tag{4.10}$$

where T_{dwell} is the dwell time on each spot and is given by

$$T_{dwell} = T_R + 2\frac{R}{c}. \tag{4.11}$$

In the above equation, T_R is the response time of the receiver acquisition system, R the link range, and c the velocity of light in vacuum. Using Eqs. (4.10) and (4.11), the total scan time of FOU is given by

$$T_U(\theta) \approx \frac{\pi\theta_U^2}{L_\theta^2}T_{dwell}. \tag{4.12}$$

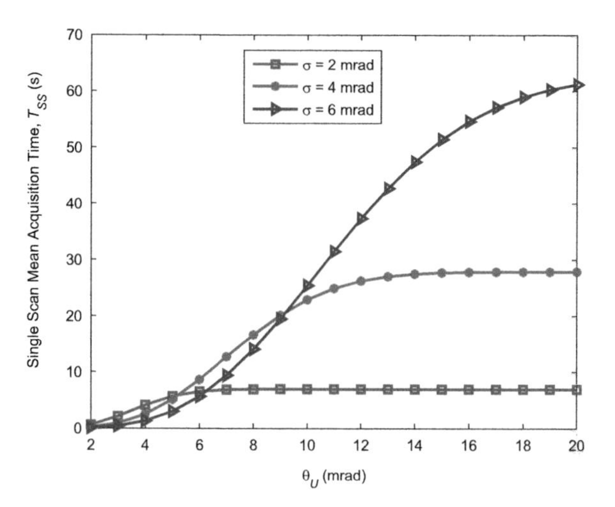

Fig. 4.7 Single-scan mean acquisition time vs. field of uncertainty

The single-scan acquisition time should be averaged with respect to the pdf of the satellite position and is given by

$$T_{SS} = \int_0^{\theta_U} T(\theta) f(\theta)\, d\theta. \tag{4.13}$$

Substituting Eqs. (4.7) and (4.12) in Eq. (4.13) gives

$$T_{SS} = \frac{2\pi\sigma_\theta^2}{L_\theta^2}\left[1 - \left(\frac{\theta_U^2}{2\sigma_\theta^2} + 1\right)\exp\left(\frac{\theta_U^2}{2\sigma_\theta^2}\right)\right] T_{dwell}. \tag{4.14}$$

This equation shows that the acquisition time is a function of dwell time (T_{dwell}), step length of the scan (L_θ), pointing error (σ_θ), and FOU (θ_U). The variation of acquisition time with FOU is shown in Fig. 4.7.

(ii) Step spiral scan: This is a variant of continuous spiral scan where the beam turns sharp corners as it scans over the uncertainty region. In this case also, the step length (L_θ) is determined by beam divergence angle, overlap factor, and uncertainty area. Just like continuous spiral scan, it also requires constant linear voltage as it scans outwards making small steps over the uncertainty region.

(iii) Segmented scan: In this case, the scans are broken into segments or sections of total uncertainty area. The center of the uncertainty area where the probability of target detection is the highest is scanned first to reduce the acquisition time. The scanning pattern is shown in Fig. 4.8.

(iv) Raster scan: In this approach, the uncertainty area is scanned at the edges, and it thus increases the acquisition time. The raster scan involves scanning one axis over the full range and incrementing the other axis at the end of field.

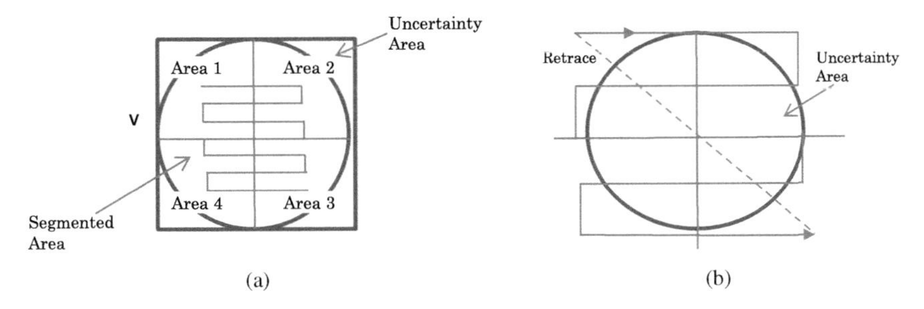

Fig. 4.8 Segmented and raster scan, (**a**) segmented scan, and (**b**) raster scan

After the complete scan of the uncertainty area, the scan beam is returned to the same position from where it was initiated as shown in Fig. 4.8. This scanning technique is less efficient than spiral scan.

4.1.3 Acquisition Approach

The acquisition process can be accomplished by using several approaches, viz., stare/stare, stare/scan, scan/scan, and scan/stare. These are briefly discussed as follows:

(i) Stare/stare approach
In this approach, the laser beam divergence is of sufficient size to illuminate the entire uncertainty area. At the same time, the receiver FOV is large enough to view the entire uncertainty area. Acquisition occurs instantaneously with acquisition probability equal to the product of probability of detection and statistical area coverage, i.e., $P_{acq} = P_{det}P_{area}$. This approach is suitable for short distances as sufficient laser power is not available to allow the transmit beam to be stared over the uncertainty area. Hence this technique is not suited for ground-to-satellite communication.

(ii) Stare/scan approach
This approach involves staring the receiver FOV and scanning a narrow laser beam over the uncertainty area. Here, one terminal (the initiating terminal, say, terminal A) slowly scans the initial uncertainty region by transmitting the beacon signal and, at the same time, the target terminal (say terminal B) searches over its FOV for the transmitted beacon signal. Once the beacon signal is detected, the target terminal then transmits a return signal to the initiating terminal which, upon detecting the return link, stops its acquisition scan. This process is illustrated in Fig. 4.9. Since a narrow laser beam is used to illuminate the uncertainty area, it allows sufficient laser power to enable detectability of the signal. Acquisition probability is the same as in the stare/stare technique,

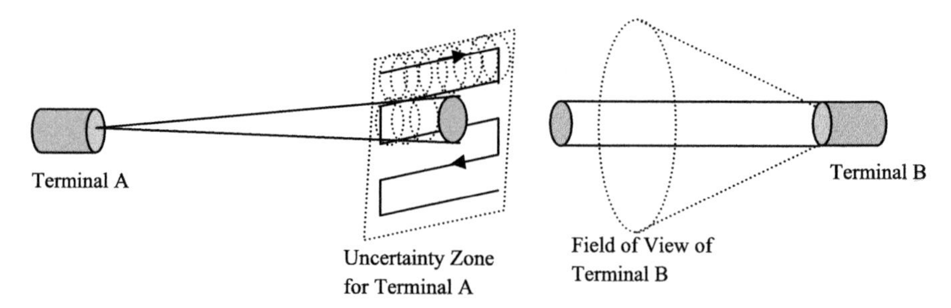

Fig. 4.9 Stare/scan acquisition technique where one terminal (*Terminal A*) slowly scans its transmitting signal while other terminal (*Terminal B*) scans through its entire uncertain region

but acquisition does not occur instantaneously. The time to acquire target is approximately given by [8]

$$t_{acq(stare/scan)} \approx \left[\frac{\theta_{unc}^2}{\theta_{div}^2 \xi_t} \right] T_{dwell} N_t, \tag{4.15}$$

where θ_{unc} is the uncertainty area in terms of solid angle, θ_{div} the beam divergence of the transmit laser source exciting the aperture, T_{dwell} the time transmitter dwells in any location, and N_t the number of total transmitter scan area repeats. The parameter ξ_t is the factor that the acquisition designer adds to the beam to provide some additional safety margin against high-frequency jitter fluctuations. This safety margin can be provided by adding some overlap to the beam on the order of scan rate. With the overlap of 10–15 % as a general thumb rule, ξ_t is given as

$$\xi_t = (1 - \epsilon)^2, \tag{4.16}$$

where ε is the overlap factor which is ≤ 1.

(iii) Scan/scan approach

In the scan/scan approach, both transmitter beam and receiver FOV are scanned simultaneously. This technique is also called parallel acquisition scanning since both the terminals can transmit a beacon signal simultaneously that rapidly scan through the uncertainty region. This approach is suitable if only one terminal has small initial pointing uncertainty. This technique is however used sparingly since the additional requirement to scan the receiver FOV increases the acquisition time by the ratio of uncertainty area/solid angle to the FOV. Acquisition probability remains the same as in the stare/stare approach. Acquisition time in this case is given as [8]

$$t_{acq(scan/scan)} \approx \left[\frac{\theta_{unc}^2}{\theta_{div}^2 \xi_t} \right] T_{dwell} N_t \left[\frac{\theta_{unc}^2}{\theta_{FOV}^2 \xi_r} \right] R_{dwell} N_r, \tag{4.17}$$

where θ_{unc}, θ_{div}, T_{dwell}, N_t, and ξ_t are the same as in the stare/scan approach discussed above. θ_{FOV} is the angular FOV of the receiver, ξ_r the receiver FOV scan overlap, R_{dwell} the receiver scan dwell time, and N_r the number of receiver scan repeats.

(iv) Scan/stare approach

A very seldom used technique, i.e., scan/stare involves scanning the receiver FOV and staring the transmitter. Once again, wide beam divergence does not provide sufficient optical energy at the receiver. This makes it unsuitable for long-distance communication. In this case, the acquisition time is given by [8]

$$t_{acq(scan/stare)} \approx \left[\frac{\theta_{unc}^2}{\theta_{FOV}^2 \xi_r}\right] R_{dwell} N_r, \tag{4.18}$$

where all the parameters are the same as used in the scan/scan approach.

From the above description, it is concluded that the stare/scan technique is well suited for ground-to-satellite uplink as its transmitter scan provides sufficient detectable power at the receiver.

4.1.4 Beam Divergence and Power Criteria for Acquisition

The acquisition process consists of two steps. First, there must be adequate signal level for initial detection at the distant spaceborne satellite and, second, sufficient received energy to allow closed-loop tracking to begin. If the beam divergence is chosen to be larger than required to acquire the target, then there must be sufficient signal energy to allow the transition to narrow beam tracking. On the other hand, small beam divergence leads to increase in acquisition time. The bound on the beam divergence comes from the required acquisition time and transition power margin. Therefore, the choice of beam divergence is very critical to provide enough received signal energy to support the initial detection of the target within the required acquisition time and at the same time allow the transition to narrow beam tracking [9–11]. The selected beam divergence must accommodate the required acquisition time; otherwise the requirements must be reexamined and acquisition time has to be modified. For long-distance laser communication like ground-to-satellite/satellite-to-ground uplink, a narrow beam divergence will allow sufficient power to be detected at the receiver. But at the same time, narrow beam divergence will cause the far-field irradiance profile to be located off-axis from the receiver due to slight misalignment and will result in signal loss. Also, it will increase the acquisition time as more time will be taken to scan the given uncertainty area. Hence, the proper choice of beam divergence and laser transmitted power will help improve the FSO system performance in terms of received SNR at the receiver, acquisition time, and pointing loss [12, 13].

4.2 Tracking and Pointing Requirements

The tracking and pointing errors play a very significant role in the performance of FSO communication system. Total pointing error is the sum of tracking and point ahead errors that can be modeled as Gaussian distributed random variable, and their value should be in the order of μradians to achieve the best system performance. The major contributor of tracking error is due to angular jitter in the tracking system that results from the detection process. Besides the angular jitter, there is another source of noise that includes the residual jitter with the gimbal systems, which cannot be tracked by the fast beam steering mirrors.

The ATP make use of internal pointing elements to take care of PAA. This PAA as discussed in Sect. 4.1 is used to compensate for the travel time over long distance due to relative velocity of two satellites. The feedback of angular position given by these internal pointing elements has some residual noise equivalent angle (NEA) just like a tracking system. Another source of error in pointing system is due to offset in the alignment between transmitter and receiver. These errors generated due to PAA and alignment offsets are generally of very small magnitude. Total pointing error in ATP system is shown in Fig. 4.10.

Most of the FSO communication system uses quadrant detectors (detector divided into four equal sectors), i.e., quadrant avalanche photodetectors (QAPD) or quadrant P-intrinsic (QPIN) photodetectors. The azimuth and elevation tracking signals (as shown in Fig. 4.11) produced by the quadrant detectors are given by

$$\text{Azimuth signal} = \frac{A + B - C - D}{A + B + C + D} \tag{4.19}$$

and

$$\text{Elevation signal} = \frac{A + C - B - D}{A + B + C + D}, \tag{4.20}$$

where A, B, C, and D are the quadrant signal levels.

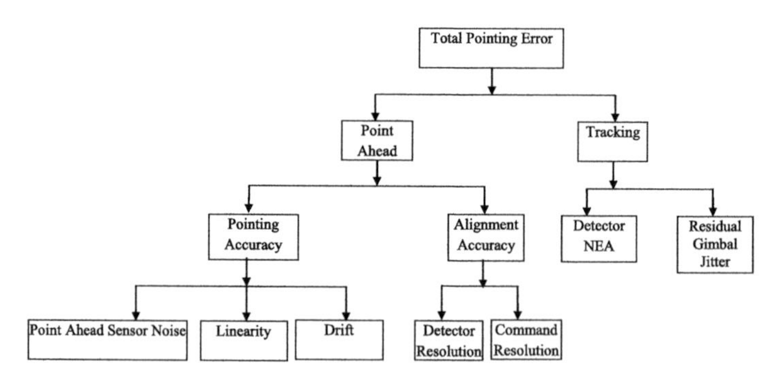

Fig. 4.10 Total pointing error

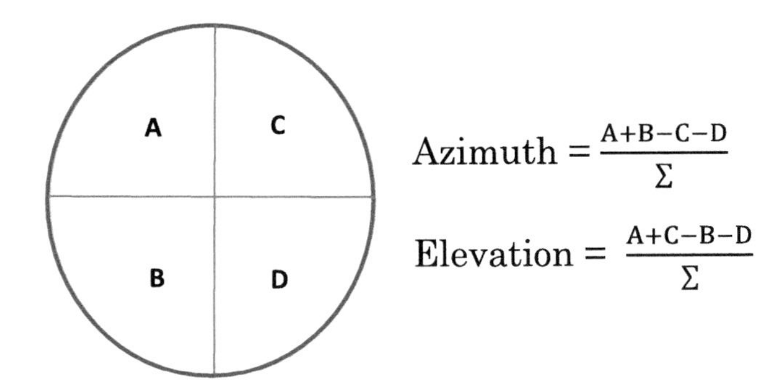

Fig. 4.11 Quadrant detector

Equations (4.19) and (4.20) represent azimuth and elevation signals when no noise signal is present. However, when noise in the tracking system is present, these equations are modified as follows

$$\text{Azimuth signal} = \frac{(A + N_A) + (B + N_B) - (C + Nc) - (D + N_D)}{A + B + C + D + N_A + N_B + N_C + N_D} \tag{4.21}$$

and

$$\text{Elevation signal} = \frac{(A + N_A) + (C + N_C) - (B + N_D) - (D + N_D)}{A + B + C + D + N_A + N_B + N_C + N_D}. \tag{4.22}$$

In the above equations, N_A, N_B, N_C, and N_D are the quadrant noise signal levels. For a very large SNR, the above expressions can be rewritten as [8]

$$\text{Azimuth signal (AZ)} = f(x) + \frac{N_A + N_B - N_C - N_D}{\Sigma} \tag{4.23}$$

and

$$\text{Elevation signal } (EL) = f(y) + \frac{N_A + N_C - N_B - N_D}{\Sigma}, \tag{4.24}$$

where Σ is the sum of A, B, C, and D, and $f(x)$ and $f(y)$ are the quadrant difference signal given by

$$f(x) = \frac{(A + B - C - D)}{\Sigma} \tag{4.25}$$

and

$$f(y) = \frac{A + C - B - D}{\Sigma}. \tag{4.26}$$

Assuming zero mean for every noise term in Eqs. (4.23) and (4.24), the expected value of azimuth and elevation signals are as follows:

$$E\,[\mathrm{AZ}] = f\,(x) \tag{4.27}$$

and

$$E\,[\mathrm{EL}] = f\,(y)\,, \tag{4.28}$$

where $E[\,]$ denotes the expected value. Therefore, the variance of Eqs. (4.27) and (4.28) is given as

$$\sigma_{AZ}^2 = E\left[\mathrm{AZ}^2\right] - \left(E\,[\mathrm{AZ}]\right)^2, \tag{4.29}$$

$$\sigma_{AZ}^2 = E\,[f\,(x)\cdot f\,(x)] + \frac{E\left[N_A^2\right] + E\left[N_B^2\right] + E\left(N_C^2\right) + E\left(N_D^2\right)}{\Sigma^2} - E\,[f\,(x)]\cdot E\,[f\,(x)]\,. \tag{4.30}$$

Since $f\,(x)$ is statistically independent, therefore,

$$\sigma_{AZ}^2 = \frac{E\left[N_A^2\right] + E\left[N_B^2\right] + E\left(N_C^2\right) + E\left(N_D^2\right)}{\Sigma^2}. \tag{4.31}$$

Since the expected value of noise terms are ideally their variances, Eqs. (4.30) and (4.31) can be rewritten as

$$\sigma_{AZ}^2 = \frac{\sigma_A^2 + \sigma_B^2 + \sigma_C^2 + \sigma_C^2}{\Sigma^2}. \tag{4.32}$$

Assuming all noise variances are equal, Eq. (4.32) can be rewritten as

$$\sigma_{AZ}^2 = \frac{4\sigma_N^2}{\Sigma^2}, \tag{4.33}$$

where σ_N^2 is the noise variance in each quadrant. Similarly variance of elevation angle is given by

$$\sigma_{EL}^2 = \frac{4\sigma_N^2}{\Sigma^2} \tag{4.34}$$

It is seen from Eqs. (4.33) and (4.34) that the numerators are mean square total noise power and denominators are mean square total signal power; therefore its mean square value can be written as inverse of SNR as

$$\sigma_{track-rms}^2 \approx \frac{1}{\mathrm{SNR}} \tag{4.35}$$

or rms value can be written as

$$\sigma_{track-rms} = \frac{1}{\sqrt{\text{SNR}}} \tag{4.36}$$

The SNR in the above expression is the signal-to-noise ratio in tracking signal bandwidth. The value of $\sigma_{track-rms}$ is also referred to as NEA of the tracking system. This RMS noise voltage can be converted to angular coordinates that requires multiplying Eq. (4.36) by voltage to angle transfer function and is given as [8]

$$\sigma_{track-rms} = \frac{1}{SF \cdot \sqrt{SNR}} \tag{4.37}$$

where SF is the angular slope factor of the tracking system that converts the angular offset to a linear voltage. The slope factor has been evaluated for Airy and Gaussian intensity profiles. For Airy and Gaussian intensity pattern, it is expressed as

$$SF = \begin{cases} 4.14/\theta_S & \text{for Airy profile} \\ 1.56/\theta_S & \text{for Gaussian profile,} \end{cases} \tag{4.38}$$

where θ_S is the optical source diameter in radian.

4.3 Integration of Complete ATP System

For establishing a link between ground station and onboard satellite, an ATP system is required to acquire and track the beacon signal with the help of several optics mechanism, optical/mechanical sensors, and beam steering mirrors so that the pointing accuracy of sub-microradian level is achieved. In order to have an accurate LOS link between ground-based station and the onboard satellite, one must provide some pointing reference. This pointing reference can be either an optical downlink beam or a beacon uplink signal. Before the actual data transmission, the ATP system first sends a reference beacon signal to scan the uncertainty region. The beacon signal should have sufficient peak power and low pulse rate to help the target receiver in locating the beam in the presence of atmospheric turbulence and large background radiation. The coarse detection is done during the acquisition mode with a large FOV. The target receiver also searches over its FOV for the beacon signal. Once the uplink beacon signal is detected, the target receiver transmits the return signal to the ground station for the transition from wide acquisition beam to narrow tracking and pointing beam. This is accomplished with the help of pointing and beam steering mirrors. Figure 4.12 shows the basic concept of the ATP system. The uplink signal from the ground station is collected with the help of telescope optics and given to the dichroic beam splitter 1. This beam splitter will reflect all the incoming signal to the beam splitter 2 that will further direct the signal to the

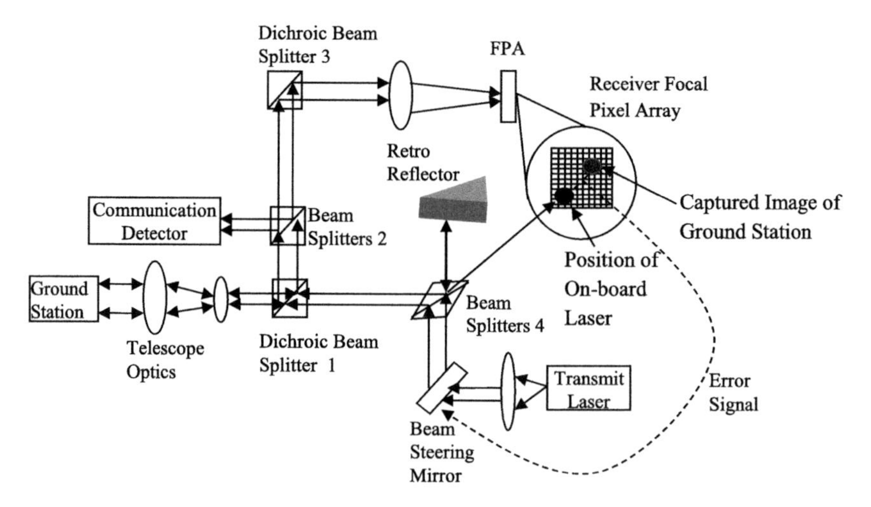

Fig. 4.12 Block diagram of ATP system between ground station and onboard satellite

communication detector or ATP subsystem in accordance with the wavelength of the incoming signal. In the case of beacon signal, the beam splitter 3 will help in focusing the image of the ground station to a point on the FPA. The location of this point on the array represents the direction from the received beacon signal relative to the telescope's axis (the center of the array). The size of the FPA determines the FOV of the telescope and is large enough to accommodate the initial pointing uncertainties. Up to this point, the position of the ground station relative to the space craft is captured on the FPA. Next step is to direct the beam from laser mounted onboard satellite to the ground-based receiver by forming a LOS link. For this, the angle of beam steering mirror has to be controlled at which the transmitted and received signals are aligned. This angular control of beam steering mirror is achieved by the error signal that is the difference between the current laser position mounted onboard satellite and the beacon signal from the ground station. The error signal will keep on driving the beam steering mirror till it reaches to a minimum value which implies that both the onboard and ground station lasers are now aligned to each other.

4.4 ATP Link Budget

For estimating the link budget of the ATP system, we need to know the required signal power that depends upon atmospheric losses and various noise sources. For acquisition and tracking system, major noise contribution is due to background noise, preamplifier noise, shot noise, thermal noise, and beat noise components due to ASE of the amplifier (i.e., ASE-ASE beat noise, ASE-signal beat noise, ASE-background beat noise). These noise sources are discussed in Chap. 3. The optical power received, P_R, by the receiver can be obtained by link range equation and is given by

$$P_R = P_T \eta_T \eta_R \left(\frac{\lambda}{4\pi R}\right)^2 G_T G_R L_T(\theta_T) L_R(\theta_R), \qquad (4.39)$$

where P_T is the transmitted optical power, η_T and η_R are the transmitter and receiver optics efficiencies, respectively, λ the operating wavelength, R the link range, G_T and G_R are the transmitter and receiver gains, respectively, and $L_T(\theta_T)$ and $L_R(\theta_R)$ are the transmitter and receiver pointing loss factors, respectively. The pointing loss factor is due to misalignment between transmitter and receiver. For Gaussian beam, the pointing loss factor is expressed as

$$L_T(\theta_T) = \exp\left(-G_T \theta_T^2\right) \quad \text{for transmitter} \qquad (4.40)$$

and

$$L_R(\theta_R) = \exp\left(-G_R \theta_R^2\right) \quad \text{for receiver.} \qquad (4.41)$$

Table 4.1 summarizes the typical link budget allocation for acquisition, tracking, and communication modes in GEO-to-GEO cross-link [14].

The link margin of acquisition link is lesser as compared to tracking and communication links. This link margin may vary depending upon design parameters and choice of components.

Table 4.1 Acquisition, tracking, and communication link margin (2.5 Gbps, DPSK modulation, BER of 10^{-9} with 5 dB coding gain at 1550 nm wavelength)

Parameter	Acquisition link	Tracking link	Communication link
Average laser power (dBW)	+7.0 (5000 mW)	+7.0 (5000 mW)	+7.0 (5000 mW)
Transmitter optics loss (dB)	−2.5	−3.0	−3.0
Transmitter gain $(\pi D_T/\lambda)^2$(dB)	+89.7 (1.5 cm)	+112.15 (15 cm)	+112.15 (15 cm)
Defocusing/truncation loss (dB)	−0.9 (200 μrad)	−0.9 (14.3 μrad)	−0.9 (14.3 μrad)
Pointing and tracking loss (dB)	−4.5	−3.0	−1.5
Range loss$(\lambda/4\pi R)^2$(dB)	−296.76 (85,000 km)	−296.76 (85,000 km)	−296.76 (85,000 km)
Receiver gain $(\pi D_R/\lambda)^2$	+112.15 (15 cm)	+112.15 (15 cm)	+112.15 (15 cm)
Receiver optics loss (dB)	−10	−2.5	−2.5
Received signal (dBW)	−105.81	−74.86	−73.36
Required signal (dBW)	−110.0 (10 pW)	−83.0 (5 nW)	−83.0 (5 nW)
Link margin (dB)	+4.19	+8.1	+9.64

4.5 Summary

Due to narrow beam of optical signal, a very tight acquisition, tracking, and pointing requirement is essential for a reliable FSO communication system. Inaccurate beam pointing and tracking can result in loss of data or large signal fades at the receiver leading to degradation of system performance. Before the actual transmission of data, spatial acquisition of the onboard satellite within a specified time limit using a narrow laser beam is a critical task. The acquisition link for the FSO system is described by acquisition time, uncertainty area, scanning techniques, and acquisition approaches that have been discussed in this chapter. Acquisition time can be traded for beam divergence and power, and this understanding is very important for designing ATP links. Errors due to pointing and tracking can significantly affect the performance of the FSO communication system. Tracking and pointing errors of ATP system have been discussed, and a complete ATP link budget has been presented. This analysis will help design engineers for a good choice of components and better system understanding.

Bibliography

1. A.A. Portillo, G.G. Ortiz, C. Racho, Fine pointing control for free-space optical communication, in *Proceeding of IEEE Conference on Aerospace*, Piscataway, vol. 3, 2001, pp. 1541–1550
2. D. Giggenbach, Lasercomm activities at the German aerospace center's institute of communications and navigation, in *International Conference on Space Optical Systems and Applications*, Corsica (2012)
3. M. Katzman, *Laser Satellite Communications* (Prentice Hall Inc., Englewood Cliffs, 1987)
4. D.J. Davis, R.B. Deadrick, J.R. Stahlman, ALGS: design and testing of an earth-to-satellite optical transceiver. Proc. SPIE-Free-Sp. Laser Commun. Technol.-V **1866**(107), 107 (1993)
5. S. Bloom, E. Korevaar, J. Schuster, H. Willebrand, Understanding the performance of free-space optics. J. Opt. Netw **2**(6), 178–200 (2003)
6. M. Toyoshima, T. Jono, K. Nakagawa, A. Yamamoto, Optimum divergence angle of a Gaussian beam wave in the presence of random jitter in free-space laser communication systems. J. Opt. Soc. Am. **19**, 567–571 (2002)
7. L. Xin, Y. Siyuan, M. Jing, T. Liying, Analytical expression and optimization of spatial acquisition for intersatellite optical communications. Opt. Exp. **19**(3), 2381–2390 (2011)
8. S.G. Lambert, W.L. Casey, *Laser Communications in Space* (Artech House Inc., Norwood, 1995)
9. T. Flom, Spaceborne laser radar. Appl. Opt. **11**(2), 291–299 (1972)
10. I.M. Teplyakov, Acquisition and tracking of laser beams in space communications. Acta Astronaut. **7**(3), 341–355 (1980)
11. J.M. Lopez, K.Yong, Acquisition, tracking, and fine pointing control of space-based laser communication systems. *Proceedings of SPIE-Control and Communication Technology in Laser Systems*, Bellingham, vol. 26, 1981, pp. 100–114
12. A.G. Zambrana, B.C. Vazquez, C.C. Vazquez, Asymptotic error-rate analysis of FSO links using transmit laser selection over gamma-gamma atmospheric turbulence channels with pointing errors. J. Opt. Exp. **20**(3), 2096–2109 (2012)

13. D.K. Borah, D.G. Voelz, Pointing error effects on free-space optical communication links in the presence of atmospheric turbulence. J. Lightwave Technol. **27**(18), 3965–3973 (2009)
14. H. Hemmati, *Near-Earth Laser Communications* (CRC Press/Taylor & Francis Group, Boca Raton, 2009)

Chapter 5
BER Performance of FSO System

5.1 System Model

The FSO communication system model consists of three main functional elements: transmitter, atmospheric channel, and receiver. The transmitter transmits the optical signal toward the receiver through atmospheric channel. The received optical signal is converted to electrical photocurrent using photodetector at the receiver. The received photocurrent can be well approximated as

$$y(t) = \eta x(t) \otimes h(t) + n(t), \tag{5.1}$$

where $\otimes$ denotes convolution, η the detector efficiency, $x(t)$ the instantaneous transmitted optical power, $h(t)$ the channel response, and $n(t)$ the additive white Gaussian noise (AWGN) with zero mean and variance σ_n^2.

The receiver can employ multiple receiving antennae in order to mitigate the effect of atmospheric turbulence. Figure 5.1 shows the plot of probability density function (pdf) of received irradiance for multiple receiver antennae ($M = 1, 3, 7,$ and 10). It is seen that with the increase in the number of antennae, peak of the profile increases and it shifts toward right.

5.2 BER Evaluation

In this section, performance in terms of BER is evaluated for coherent (SC-BPSK and SC-QPSK) and noncoherent (OOK and $\mathbb{M}$-PPM) schemes in the presence of weak atmospheric turbulence. In the optical receiver, the BER is functionally related to the received signal irradiance I. In FSO uplink, this irradiance I fluctuates randomly due to turbulence in the atmosphere. The statistical variation of I depends upon the turbulence level. As discussed earlier, the pdf of I can be lognormal,

H. Kaushal et al., *Free Space Optical Communication*, Optical Networks,
DOI 10.1007/978-81-322-3691-7_5

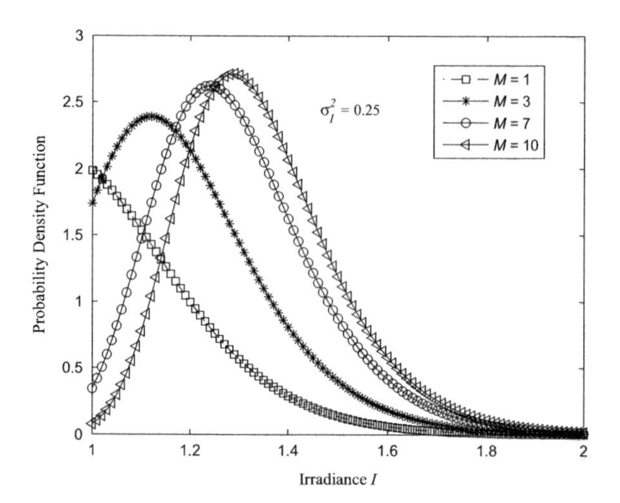

Fig. 5.1 The received irradiance pdf for various values of receiver antennae ($M = 1, 3, 7,$ and 10) in weak atmospheric turbulence level of $\sigma_I^2 = 0.25$

exponential, K distribution, etc. In weak atmospheric turbulence, random variable I is characterized by lognormal distribution. The BER in the presence of turbulence is given by

$$P_e = Q\left(\sqrt{SNR\left(I\right)}\right). \tag{5.2}$$

The unconditional BER is obtained by averaging Eq. (5.2) over irradiance fluctuations statistics and is given as

$$P_e = \int_0^{\infty} f_I\left(I\right) Q\left(\sqrt{SNR\left(I\right)}\right) dI. \tag{5.3}$$

As $f_I\left(I\right)$ is lognormally distributed [1], the above equation can be written as

$$P_e = \int_0^{\infty} \frac{1}{I\sqrt{2\pi\sigma_I^2}} \exp\left\{-\frac{\left(\ln\left(I/I_0\right) + \sigma_I^2/2\right)^2}{2\sigma_I^2}\right\} Q\left(\sqrt{SNR\left(I\right)}\right) dI. \tag{5.4}$$

The SNR without turbulence is given by $SNR = \left(R_0 A I\right)^2 / 2\sigma_n^2$ where I is the received irradiance, R_0 the photodetector responsivity, and A the photodetector area which is taken as unity so that the received optical power is the same as the received signal irradiance. A closed-form solution of Eq. (5.4) is not possible. However, by using an alternative representation of Q-function [2, 3] together with the Gauss-Hermite quadrature integration, the above equation can be rewritten as

$$P_e \cong \int_{-\infty}^{\infty} \frac{1}{\sqrt{\pi}} \exp\left(-x^2\right) Q\left(\frac{R_0 A I_0 \exp\left(\sqrt{2}\sigma_I x - \sigma_I^2/2\right)}{\sqrt{2}\sigma_n}\right) dx, \tag{5.5}$$

where $x = \left(\frac{\ln(I/I_0)+\sigma_I^2/2}{\sqrt{2}\sigma_I}\right)$. The above equation can be approximated using Gauss-Hermite quadrature integration as given below

$$\int_{-\infty}^{\infty} f(x) \exp\left(-x^2\right) dx \cong \sum_{i=1}^{m} w_i f(x_i), \tag{5.6}$$

where x_i and w_i represent the zeros and weights of the mth order Hermite polynomials. The degree of accuracy of above equation depends upon the order of Hermite polynomial.

5.2.1 Coherent Subcarrier Modulation Schemes

In a coherent subcarrier modulation scheme, RF signal (subcarrier) is pre-modulated with the information signal and is then used to modulate the intensity of the optical carrier. In case of subcarrier BPSK, the RF subcarrier at frequency f_c is modulated in accordance with the information signal in which bits "1" and "0" are represented by two different phases that are 180° apart. Here, the RF subcarrier is represented by

$$s(t) = A_c \cos(\omega_c t + \phi), \tag{5.7}$$

where A_c is the peak amplitude taken to be unity and ω_c and ϕ are the subcarrier angular frequency and phase, respectively. The phase ϕ can take the values 0 or π depending upon the information bits. Since the subcarrier signal $s(t)$ is sinusoidal having both positive and negative values, a DC level is added to modulate it onto the intensity of an optical carrier. This will ensure that the bias current is always equal to or greater than the threshold current. Hence, the modulating signal to the optical source in the transmitter is given by

$$m(t) = 1 + \beta \cos(\omega_c t + \phi), \tag{5.8}$$

where β is the modulation index ($\beta \leq 1$). The electrical signal at the photodetector output will be

$$I_D(t) = R_0 A I \left[1 + \beta\left(\cos(\omega_c t + \phi)\right)\right] + n(t), \tag{5.9}$$

where $n(t)$ is the AWGN with zero mean and variance σ_n^2. The other parameters R_0, A, and I are as defined earlier. For coherent demodulation, the above received signal is multiplied by a locally generated RF subcarrier signal at the same frequency ω_c and then pass through a LPF. Therefore, the signal at the filter output is given by

$$I_r(t) = \frac{R_0 A I \beta \cos\phi}{2} + \frac{n_I(t)}{2}, \tag{5.10}$$

where $n_I(t)$ is the in-phase noise component. The probability of error in case of SC-BPSK is given by

$$P_e = P(0)\,P(I_r > Th/0) + P(1)\,P(I_r < Th/1)\,, \tag{5.11}$$

where $P(0)$ and $P(1)$ are the probability of transmitting data bits "0" and "1," respectively. Assuming equiprobable data transmission such that $P(0) = P(1) = 0.5$, the above equation becomes

$$P_e = 0.5\left[P(I_r > Th/0) + P(I_r < Th/1)\right]. \tag{5.12}$$

The marginal probabilities $P(I_r > Th/0)$ and $P(I_r < Th/1)$ are given by

$$P(I_r > Th/0) = \int_{Th}^{\infty} \frac{1}{\sqrt{2\pi\sigma_n^2}} \exp\left[-\frac{(I_r - I_m)^2}{2\sigma_n^2}\right] dI_r \tag{5.13}$$

and

$$P(I_r < Th/1) = \int_{-\infty}^{Th} \frac{1}{\sqrt{2\pi\sigma_n^2}} \exp\left[-\frac{(I_r + I_m)^2}{2\sigma_n^2}\right] dI_r, \tag{5.14}$$

where $I_m = R_0AI/2$ for unity modulation index, i.e., $\beta = 1$. Corresponding to transmitted bits "1" and "0," signal components of the current from Eq. (5.10) are I_m ($\phi = 0$) and $-I_m$ ($\phi = 180°$), respectively, and therefore the decision threshold Th is chosen to be zero. From Eqs. (5.12), (5.13), and (5.14), the BER can be written as

$$P_e = \int_{0}^{\infty} \frac{1}{\sqrt{2\pi\sigma_n^2}} \exp\left[-\frac{(I_r - I_m)^2}{2\sigma_n^2}\right] dI_r \tag{5.15}$$

$$= \frac{1}{2}\operatorname{erfc}\left(\frac{I_m}{\sigma_n}\right) = Q\left(\frac{\sqrt{2}I_m}{\sigma_n}\right) = Q\left(\sqrt{SNR}\right), \tag{5.16}$$

where $\sqrt{SNR} = R_0AI/\sqrt{2}\sigma_n$. In presence of atmospheric turbulence, the conditional averaging over irradiance is performed, and P_e(SC-BPSK) is given as

$$P_e(\text{SC} - \text{BPSK}) = \int_{0}^{\infty} Q\left(\sqrt{SNR}\right) f_I(I)\, dI. \tag{5.17}$$

The above equation can be approximated by Gauss-Hermite quadrature integration (refer to Eqs. (5.4), (5.5), and (5.6)). Further, with the change of variable,

Table 5.1 Values of K_a and K_b for different noise-limiting conditions

Parameter	Quantum limit	Thermal noise	Background noise	Thermal and background noise
K_a	$\frac{\beta^2 R_0 I_0 P_{av}}{2qR_b}$	$\frac{(\beta R_0 I_0)^2 P_{av} R_L}{4KTR_b}$	$\frac{(\beta I_0)^2 R_0 P_{av}}{2qR_b I_{BG}}$	$\frac{(\beta R_0 I_0)^2 P_{av}}{\sigma_{BG}^2 + \sigma_{Th}^2}$
K_b	0.5	1	1	1

$y = \frac{\ln(I/I_0)+\sigma_I^2/2}{\sqrt{2}\sigma_I}$ in Eq. (5.4), the BER for FSO link from Eq. (5.17) is approximately given by

$$P_e\,(\text{SC}-\text{BPSK}) \cong \frac{1}{\sqrt{\pi}} \sum_{i=1}^{m} w_i Q\left(K_a \exp\left[\left(K_b \sigma_I x_i - \sigma_I^2/2\right)\right]\right). \tag{5.18}$$

Similarly, the BER for SC-QPSK following [4, 5] will be

$$P_e\,(\text{SC}-\text{QPSK}) \cong \frac{1}{\sqrt{\pi}} \sum_{i=1}^{m} w_i Q\left(\frac{K_a \exp\left[\left(K_b \sigma_I x_i - \sigma_I^2/2\right)\right]}{\sqrt{2}}\right). \tag{5.19}$$

The values of K_a and K_b are given in Table 5.1 for different noise-limiting conditions.

In Table 5.1, P_{av} is average power given by $P_{av} =_{Lt\ T\to\infty} \frac{1}{T}\int_0^T s^2(t)\,dt$, R_b is the data rate, and σ_{Bg}^2 and σ_{Th}^2 are the variance due to background and thermal noise components, respectively. The background noise due to extended and point sources is a shot noise and its variance is given by [6]

$$\sigma_{Bg}^2 = 2qR_0 P_B B = 2qR_0\left(I_{sky} + I_{sun}\right) A_r B. \tag{5.20}$$

In the above equation, A_r is the receiver area and I_{sky} and I_{sun} are extended and point background noise sources irradiance, respectively, given by following equations:

$$I_{sky} = N(\lambda)\,\Delta\,\lambda_{filter}\pi\,\Omega^2/4, \tag{5.21}$$

$$I_{sun} = W(\lambda)\,\Delta\lambda_{filter}, \tag{5.22}$$

where $N(\lambda)$ and $W(\lambda)$ are the spectral radiance of the sky and spectral radiance emittance of the Sun, respectively, $\Delta\lambda_{filter}$ the bandwidth of optical band pass filter (BPF) in the receiver, and Ω the FOV of the receiver in radians. The BER plot of Eqs. (5.18) and (5.19) under quantum-limited condition for weak atmospheric turbulence is shown in Fig. 5.2.

It is clear from the above figure that SC-BPSK performs better than SC-QPSK for weak atmospheric turbulence conditions. In both the cases as the level of turbulence

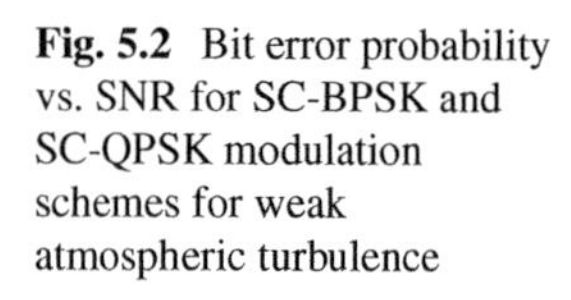

Fig. 5.2 Bit error probability vs. SNR for SC-BPSK and SC-QPSK modulation schemes for weak atmospheric turbulence

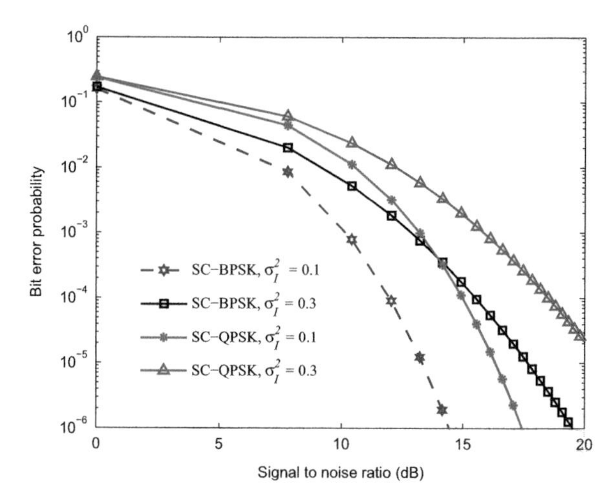

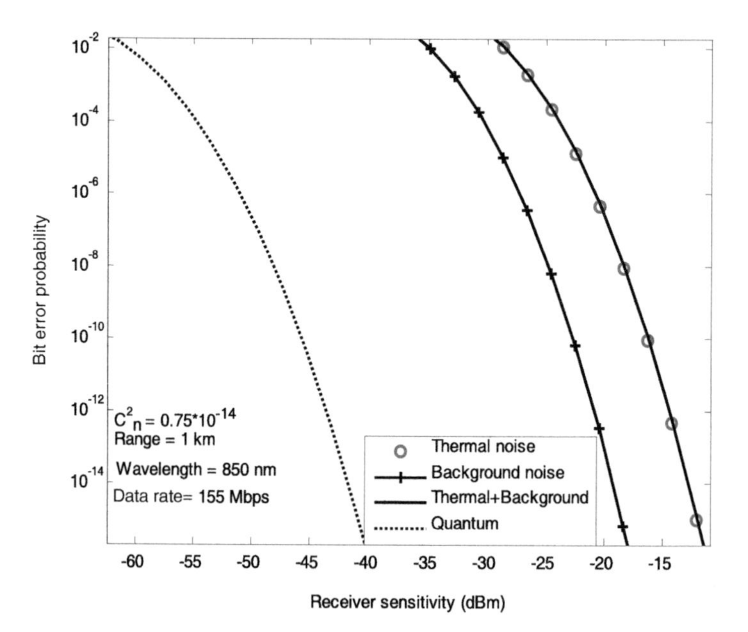

Fig. 5.3 The BER vs. receiver sensitivity for different noise sources in weak turbulence level of $\sigma_I^2 = 0.3$ [6] (This image is shared under Creative Commons Attribution 3.0 Unported License)

increases, required SNR also increases. Figure 5.3 shows the BER vs. SNR plot for different noise-limiting conditions based on the system parameters given in Table 5.2. This figure illustrates that in an FSO link, the system performance is limited by thermal noise with suitable optical BPF and narrow FOV detector. Similarly, the BER equations for other coherent modulation schemes can be derived

Table 5.2 System parameters [6]

Parameter	Value
Symbol rate, R_b	155 Mbps
Spectral sky radiance	10^{-3} W/cm^3 $\mu m Sr$
Spectral radiant emittance of the Sun, W(λ)	0.055 W/cm^2 μm
Optical BPF, $\Delta\lambda_{filter}$ at $\lambda = 850$ nm	1 nm
PIN photodetector field of view (FOV)	0.6 rad
Operating wavelength, λ	850 nm
Number of subcarrier, N	1
Link range, R	1 km
Refractive index structure parameter, C_n^2	0.75 m$^{-2/3}$
Load resistance, R_L	50 Ω
Responsivity of PIN photodetector, R_0	1 A/W
Operating temperature, T	300 K

in the presence of turbulence. The conditional BER expression for other coherent modulation schemes are given below

$$\begin{array}{ll} \mathbb{M}\text{-PSK}, \mathbb{M} \geq 4, & P_e \approx \frac{2}{\log_2 \mathbb{M}} Q\left(\sqrt{(\log_2 \mathbb{M}) \cdot SNR\,(I)} sin\left(\frac{\pi}{\mathbb{M}}\right)\right) \\ \text{DPSK}, & P_e = 0.5\exp\left(-0.5 \cdot SNR\,(I)\right) \\ \mathbb{M}\text{-QAM}, & P_e \approx \frac{4}{\log_2 \mathbb{M}} Q\left(\frac{3\log_2 \mathbb{M} \cdot SNR(I)}{\mathbb{M}-1}\right) \end{array}$$

5.2.2 *Noncoherent Modulation Schemes*

In this section, probability of error equations for OOK, $\mathbb{M}$-ary pulse-position modulation ($\mathbb{M}$-PPM), differential PPM (DPPM), differential amplitude and pulse-position modulation (DAPPM), digital pulse interval modulation (DPIM), and dual header-pulse interval modulation (DHPIM) are discussed.

5.2.2.1 On Off Keying

For OOK modulation scheme, data bit $d\,(t)$ is transmitted by pulsing the light source either “on” or “off” during each bit interval, i.e., a data bit $d\,(t) = \{0\}$ is transmitted as an absence of light pulse and $d\,(t) = \{1\}$ as a pulse of finite duration. The instantaneous photocurrent is given by

$$I_r\,(t) = R_0 A I + n\,(t)\,. \tag{5.23}$$

The BER in this case will be

$$P_e\,(\mathrm{OOK}) = P\,(0)\,P\,(I_r > Th/0) + P\,(1)\,P\,(I_r < Th/1)\,, \tag{5.24}$$

where $P\,(0)$ and $P\,(1)$ are the probabilities of transmitting data bits "0" and "1," respectively and each having the probability of transmission equal to 0.5.

$$P\,(I_r > Th/0) = \int_{Th}^{\infty} p\,(I_r/0)\,dI_r, \tag{5.25}$$

$$P\,(I_r < Th/1) = \int_{-\infty}^{Th} p\,(I_r/1)\,dI_r. \tag{5.26}$$

It is clear from Eq. (5.26) that probability of error when bit "1" is transmitted depends upon the received irradiance I_r. In weak atmospheric turbulence, the irradiance I_r follows a lognormal distribution. Therefore, Eq. (5.26) can be written as

$$\begin{aligned} P\,(I_r/1) &= \int_0^{\infty} p\,(I_r/1, I_r)\,f_{I_r}\,(I_r)\,dI_r \\ &= \int_0^{\infty} \frac{1}{\sqrt{2\pi\sigma_n^2}} \exp\left(-\frac{(I_r - Th)^2}{2\sigma_n^2}\right) \\ &\quad \cdot \frac{1}{I_r\sqrt{2\pi\sigma_l^2}} \exp\left\{-\frac{\left(\ln\left(I_r/I_0\right) + \sigma_l^2/2\right)}{2\sigma_l^2}\right\} dI_r. \end{aligned} \tag{5.27}$$

Substituting Eqs. (5.25), (5.26) and (5.27) in (5.24) and then using Eqs. (5.5) and (5.6), the BER of OOK for single-input single-output (SISO) system can be written as

$$\begin{aligned} P_e\,(\mathrm{OOK}) &= P\,(0)\,Q\,(Th/\sigma_n) \\ &+ (1 - P\,(0))\,\frac{1}{\sqrt{\pi}} \sum_{i=1}^{m} w_i Q \frac{\left[R_0 A I_0 \exp\left(\sqrt{2}\sigma_I x_i - \sigma_l^2/2\right) - Th\right]}{\sqrt{2}\sigma_n}. \end{aligned} \tag{5.28}$$

The parameter Th in the above equation is the threshold whose value is fixed midway between the two logic levels assigned for bit 1 and bit 0.

It is clear from Fig. 5.4 that for fixed value of threshold set to 0.5, the required SNR increases with the increase in the value of turbulence and almost reaches

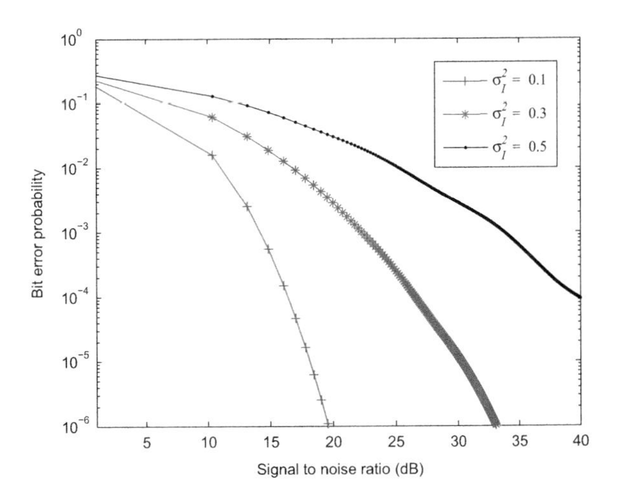

Fig. 5.4 BER vs. SNR for OOK modulation scheme in weak atmospheric turbulence

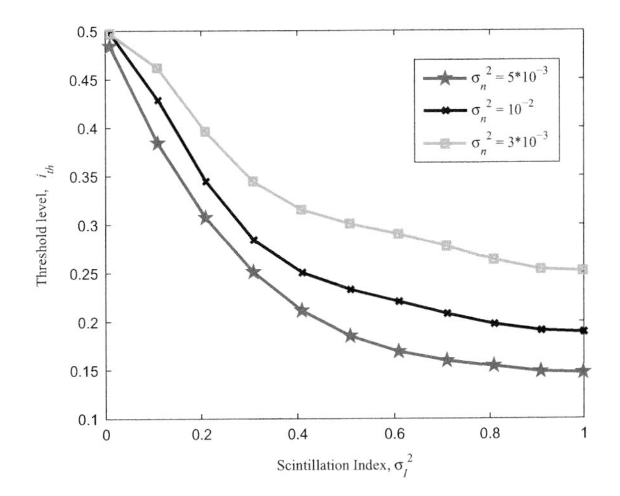

Fig. 5.5 Variation in threshold level of OOK vs. log intensity standard deviation for various noise levels [6]

to error floor at $\sigma_I^2 = 0.5$. Therefore, an adaptive threshold is required to achieve the best system performance in case of OOK modulation technique. In adaptive threshold approach, the value of threshold is varied in accordance with the turbulence in the atmosphere. Figure 5.5 shows the plots of variation in threshold with different values of atmospheric turbulence levels for OOK modulation technique at various noise levels. It is seen from this figure that if there is zero turbulence, optimum threshold level for binary signals is 0.5. However, as the value of turbulence increases, the corresponding threshold level decreases. Further, for a given atmospheric turbulence level, the threshold level increases with the increase in noise level.

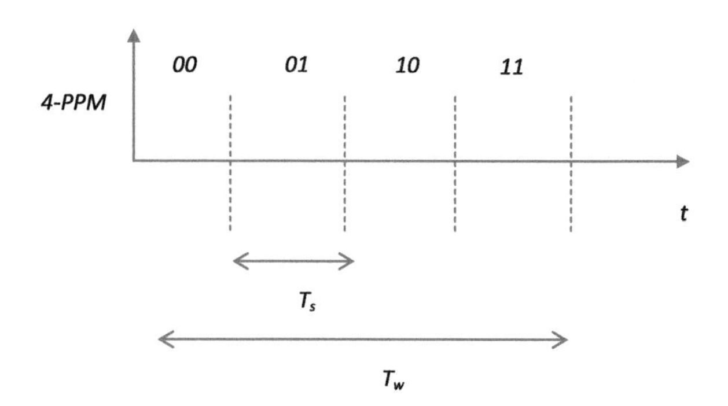

Fig. 5.6 Waveform for 4-PPM scheme

5.2.2.2 $\mathbb{M}$-ary Pulse-Position Modulation

In case of $\mathbb{M}$-PPM, each PPM symbol is divided into $\mathbb{M}$ time slots, and an optical pulse is placed in one of the $\mathbb{M}$ adjacent time slots to represent a data word. Number of bits present per PPM slot or pulse is $\log_2\mathbb{M}$. These $\mathbb{M}$ slots are called chips, and the presence and absence of pulse in the chips are called "on" and "off" chip, respectively. Therefore, one PPM word will have one "on" chip and ($\mathbb{M}$-1) "off" chips. The duration of the slot, T_s, is related to word duration, T_w, as

$$T_w = \mathbb{M}T_s. \tag{5.29}$$

Hence, the data rate for $\mathbb{M}$-PPM is given as

$$R_b = \frac{\log_2\mathbb{M}}{T_w} = \frac{\log_2\mathbb{M}}{\mathbb{M}T_s}, \tag{5.30}$$

where R_b is the data rate. Figure 5.6 shows the time slots and corresponding bit pattern for 4-PPM.

For $\mathbb{M}$-PPM, the probability of word error as a function of irradiance I can be written as

$$P_{ew}\,(\text{PPM}) = (\mathbb{M}-1)\,Q\left(\sqrt{SNR\,(I)}\right). \tag{5.31}$$

BER is related to word error probability in $\mathbb{M}$-PPM as

$$P_e\,(\mathbb{M}-\text{PPM}) = \left(\frac{\mathbb{M}/2}{\mathbb{M}-1}\right)P_{ew}. \tag{5.32}$$

Using Eqs. (5.31) and (5.32), P_e($\mathbb{M}$-PPM) can be approximated as

$$P_e\,(\mathbb{M}-\text{PPM}) = (\mathbb{M}/2)\,Q\left(\sqrt{SNR\,(I)}\right). \tag{5.33}$$

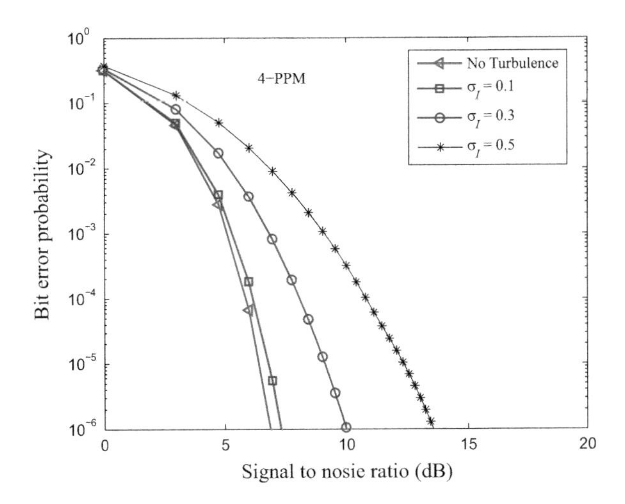

Fig. 5.7 Bit error probability vs. SNR for 4-PPM scheme in weak atmospheric turbulence

In the presence of atmospheric turbulence, the average BER for $\mathbb{M}$-PPM is given by

$$P_e(\mathbb{M} - \text{PPM}) = \int_0^{\infty} (\mathbb{M}/2)\, Q\left(\sqrt{SNR(I)}\right) f_I(I)\, dI. \tag{5.34}$$

Using Gauss-Hermite integration and substituting $f_I(I)$ as lognormal distribution in weak atmospheric turbulence, the above integration can be approximated as

$$P_e(\mathbb{M} - \text{PPM}) = \frac{\mathbb{M}}{2\sqrt{\pi}} \sum_{i=1}^{m} w_i Q\left(K\exp\left(\sqrt{2}\sigma_I x_i - \sigma_I^2/2\right)\right), \tag{5.35}$$

where $K = \left(R_0 A I_0 \mathbb{M}/\sqrt{2}\sigma_n\right)$. Figure 5.7 shows the variation of bit error probability for different values of scintillation index for 4-PPM modulation scheme.

It is always convenient to express BER in terms of photon efficiency as it is a direct way to estimate the efficiency of any system for a given error probability. Therefore, Eq. (5.34) can be rewritten as

$$P_e(\mathbb{M} - \text{PPM}) = \int_0^{\infty} (\mathbb{M}/2)\, Q\left(\sqrt{SNR(K_s)}\right) f_K(K_s)\, dK_s. \tag{5.36}$$

The expression for SNR without turbulence in terms of a number of photons is given as

$$SNR = \frac{K_s^2}{FK_s + K_n}, \tag{5.37}$$

where K_s $(= \eta\lambda P_R T_w / hc)$ is the photon count per PPM slot and F the excess noise factor of the photodetector given by

$$F = \varsigma \mathcal{M} + \left(2 - \frac{1}{\mathcal{M}}\right)(1 - \varsigma). \tag{5.38}$$

In the above equation, ς is the ionization factor and $\mathcal{M}$ the average gain of APD. The parameter K_n in Eq. (5.37) is given by

$$K_n = \left[2\sigma_n^2 / (\mathcal{M}q)^2\right] + 2FK_b, \tag{5.39}$$

where σ_n^2 is the equivalent noise count within PPM slot. In weak atmospheric turbulence, $f_K(K_s)$ is lognormally distributed as the instantaneous received irradiance is lognormally distributed. The BER expression in the presence of lognormal atmospheric turbulence can therefore be approximated as [7]

$$P_e = \frac{1}{\sqrt{\pi}} \sum_{i=1}^{m} w_i Q\left(\frac{\exp\left(2\sqrt{2}\sigma_I x_i - \sigma_I^2/2\right)}{F\exp\left(\sqrt{2}\sigma_I x_i - \sigma_I^2/2\right) + K_n}\right). \tag{5.40}$$

Using union bound, the upper bound on BER denoted by $P_e^{\mathbb{M}}$ is expressed as

$$P_e^{\mathbb{M}} \leq \frac{\mathbb{M}}{2\sqrt{\pi}} \sum_{i=1}^{m} w_i Q\left(\frac{\exp\left(2\sqrt{2}\sigma_I x_i - \sigma_I^2/2\right)}{F\exp\left(\sqrt{2}\sigma_I x_i - \sigma_I^2/2\right) + K_n}\right). \tag{5.41}$$

Figure 5.8 shows the BER plot of $\mathbb{M}$-PPM for various levels of atmospheric turbulence. It is clear from this figure that for a fixed data rate and background noise, as the value of scintillation increases, the required signal level also increases to achieve the same BER performance.

5.2.2.3 Differential PPM

In differential PPM (DPPM), also called truncated PPM (TPPM), the next new PPM symbol immediately follows the slot containing the pulse. Like in the case of $\mathbb{M}$-PPM, a block of $\log_2\mathbb{M}$ input bits are mapped on to one of the $\mathbb{M}$ distinct waveforms. Every waveform includes one "on" chip (presence of pulse) and $\mathbb{M} - 1$ "off" chips (absence of pulse). DPPM is a simple modification of $\mathbb{M}$-PPM where all "off" chips following the "on" chips are deleted. Therefore, this modulation scheme increases the bandwidth efficiency. It improves the throughput per unit time by the factor of two, since symbols are on the average half as long as they would be with ordinary PPM. However, it will lead to varying symbol size and hence imposes

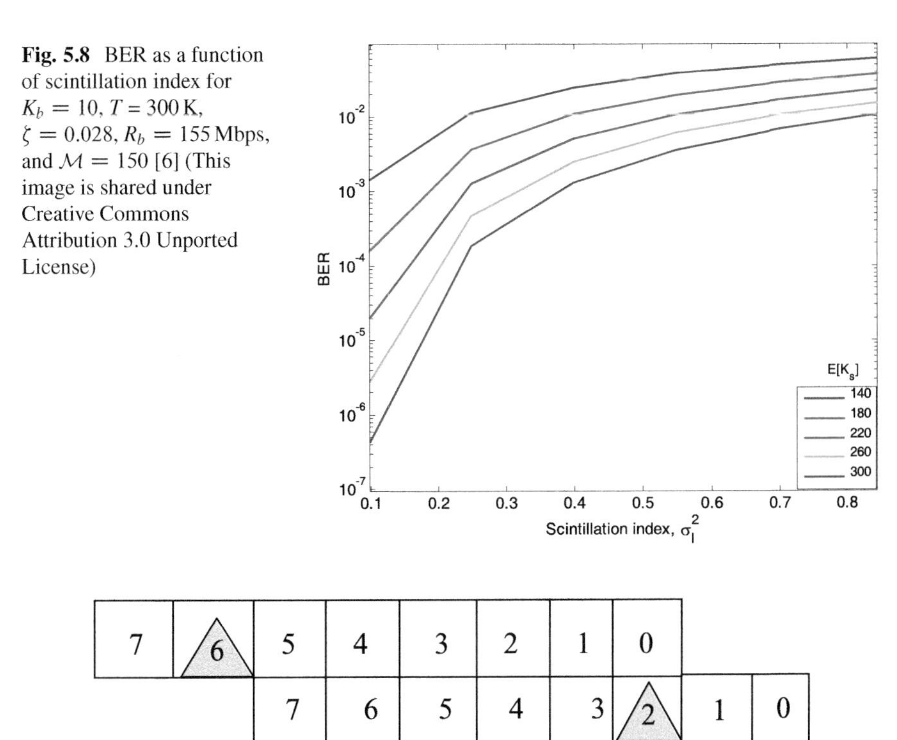

Fig. 5.8 BER as a function of scintillation index for $K_b = 10$, $T = 300$ K, $\zeta = 0.028$, $R_b = 155$ Mbps, and $\mathcal{M} = 150$ [6] (This image is shared under Creative Commons Attribution 3.0 Unported License)

Fig. 5.9 8-DPPM scheme for the transmission of message 110010

synchronization problem at the receiver. Figure 5.9 shows 8-DPPM scheme where a new symbol appears immediately after the pulse in the previous frame.

DPPM works well with Q-switched laser as it can confine very high peak power (gigawatt range) to a narrow slot. However, in *Q-switched* laser, the next pulse is delayed (also called dead time) till the time the laser is recharged again, and thus Q-switching leads to low-pulse repetition rate with longer pulse duration. PPM can cater for this dead-time constraint by following each frame by a period during which no pulses are transmitted.

A comparison of the waveforms of 4-PPM and 4-DPPM is shown in Fig. 5.10. Mapping between source bits and transmitted chips for both the modulation schemes is shown in Table 5.3. It is seen from Fig. 5.10 that in case of $\mathbb{M}$-PPM as the level of $\mathbb{M}$ is increased, the average power efficiency is improved but at the cost of reduction in bandwidth efficiency. In the case of fixed $\mathbb{M}$, $\mathbb{M}$-DPPM has a higher duty cycle and is therefore less average power efficient than $\mathbb{M}$-PPM. However, for a fixed average bit rate and available bandwidth, one can employ a higher $\mathbb{M}$ with DPPM than with PPM, resulting in a net improvement of average power efficiency [8].

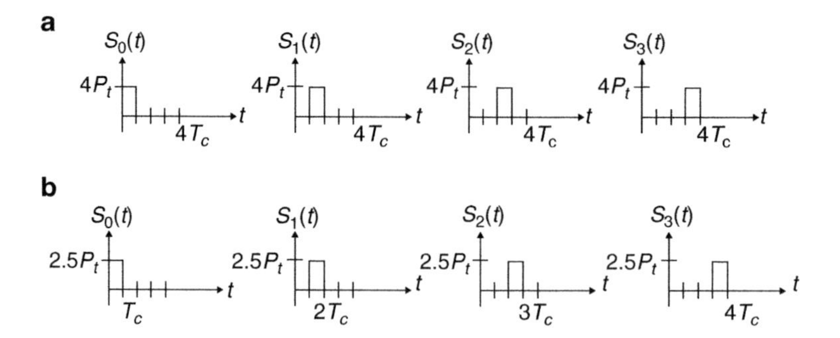

Fig. 5.10 Waveforms for (**a**) 4-PPM and (**b**) 4-DPPM using rectangular pulse. P_t is the average transmitted power and T_c is the chip duration [8]

Table 5.3 Mapping between source bits and transmitted chips of 4-PPM and 4-DPPM schemes

Source bits	4-PPM chips (nominal mapping)	4-DPPM chips (nominal mapping)	4-DPPM chips (reverse mapping)
00	1000	1	0001
01	0100	01	001
10	0010	001	01
11	0001	0001	1

The packet error probability for an X bits, N chip packets is [8] given by

$$P_{X,\text{DPPM}} = 1 - (1 - p_0)^{N - (X/\log_2 \mathbb{M})}\,(1 - p_1)^{(X/\log_2 \mathbb{M})} \approx \left(N - \frac{X}{\log_2 \mathbb{M}}\right) P(0) + \frac{X}{\log_2 \mathbb{M}} P(1), \tag{5.42}$$

where $P(0)$ is the probability that an "off" chip is detected to be "on" and vice versa for $P(1)$. If the threshold is set at the mean of expected "on" and "off" levels, i.e., $P(0) = P(1)$, then

$$P_{X,\text{DPPM}} \approx NQ\left(R_0 P_R \sqrt{\frac{(\mathbb{M}+1)\log_2 \mathbb{M}}{8 R_b N_0}}\right), \tag{5.43}$$

where N_0 is two-sided power spectral density (PSD) and R_b the average bit rate given by $R_b = \log_2 \mathbb{M} / (\mathbb{M}+1) T_s$. The bandwidth efficiency of DPPM for a given bandwidth B is given as

$$\frac{R_b}{B} = \frac{2\log_2 \mathbb{M}}{\mathbb{M}+1}. \tag{5.44}$$

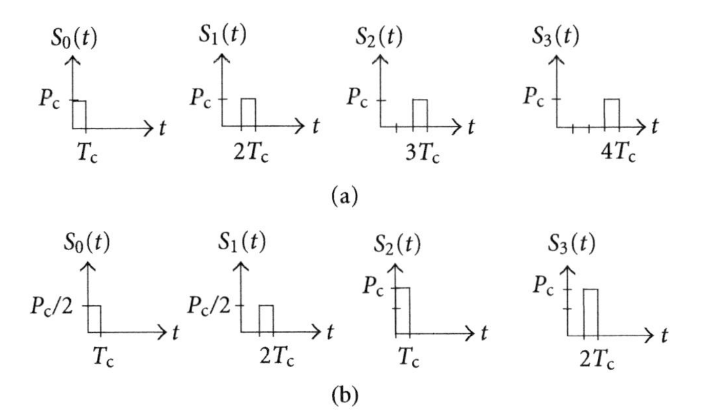

Fig. 5.11 The symbol structure for (**a**) DPPM ($\mathbb{M} = 4$) and (**b**) DAPPM ($A = 2$, $\mathbb{M} = 2$) [9]

5.2.2.4 Differential Amplitude Pulse-Position Modulation

Differential amplitude pulse-position modulation (DAPPM) is a combination of pulse amplitude modulation (PAM) and DPPM. Therefore the symbol length and pulse amplitude are varied according to the information being transmitted. Figure 5.11 gives the comparison of DPPM and DAPPM symbol structure. Here, a block of n input bits are mapped to one of the $2^n = (A \times \mathbb{M})$ distinct symbols, each of which has one "on" chip that is used to indicate the end of a symbol. The symbol length varies from $\{1,2,\ldots,\mathbb{M}\}$, and the pulse amplitude of "on" chip is selected from $\{1,2,\ldots, A\}$, where A and $\mathbb{M}$ are integers. The bit resolution is therefore given by $\log_2(A \cdot \mathbb{M})$. The average number of empty slots preceding the pulse can be lowered by increasing the number of amplitude levels A, thereby increasing the achievable throughput in the process. When compared with similar modulation techniques, a well-designed DAPPM will require the least bandwidth. DAPPM suffers from a high average power and a large DC component, thus restricting its use to applications where power is not critical. It is also susceptible to the baseline wander due to its large DC component. The packet error rate (PER) of DAPPM with maximum likelihood sequence detection (MLSD) receiver can be considered as the packet error rate of a PAM system with MLSD when the PAM symbol is $\{1, 2, \ldots .A\}$ and each symbol is equally likely and independent. The PAM packet length is equal to $D/\log_2(A \cdot \mathbb{M})$. Then, for a nondispersive channel, the packet error rate for$D$-bit packet is approximated by [9]

$$PER = 1 - (1 - P_{ce})^{\mathbb{M}_{avg}D/n} \approx \frac{\mathbb{M}_{avg}DP_{ce}}{n}, \tag{5.45}$$

where P_{ce} is the probability of chip error, $\mathbb{M}_{avg}$ the average length of a DAPPM symbol ($= \mathbb{M}+1/2$), and n the number of input bits. In case of maximum likelihood sequence detection (MLSD), packet error rate for nondispersive channels is given as [9]

$$PER = \frac{2(A-1)}{A} \frac{D}{\log_2(A \cdot \mathbb{M})} Q\left(\frac{R_0 P_R}{2A\sqrt{N_0 B}}\right). \tag{5.46}$$

The bandwidth efficiency of DAPPM is given by

$$\frac{R_b}{B} = \frac{2\log_2(A \cdot \mathbb{M})}{\mathbb{M} + A} \frac{\text{bit/s}}{\text{Hz}}. \tag{5.47}$$

Since the average length of DAPPM symbol is $\mathbb{M}_{avg} = (\mathbb{M} + 1/2)$, therefore, the average bit rate R_b is therefore given by $\log_2\mathbb{M}/(\mathbb{M}_{avg}T_c)$. The chip duration, $T_c = 2\log_2\mathbb{M}/(\mathbb{M}+1)R_b$.

Its peak-to-average power ratio is given by [9]

$$PAPR = \frac{P_c}{P_{avg}} = \frac{A(\mathbb{M}+1)}{(A+1)}. \tag{5.48}$$

5.2.2.5 Digital Pulse Interval Modulation

Digital pulse interval modulation (DPIM) is an anisochronous PPM technique in which each block of $\log_2\mathbb{M}$ data bits are mapped to one of $\mathbb{M}$ possible symbols. The symbol length is variable and is determined by the information content of the symbol. Every symbol begins with a pulse, followed by a series of$k$ empty slots, the number of which is dependent on the decimal value of the block of data bits being encoded. Therefore, for DPIM, an $\mathbb{M}$ bit symbol is represented by a pulse of constant power, P_c in "on chip" followed by k empty slots of zero power ("off chip") where $1 \le k \le 2^{\mathbb{M}}$ [10]. In order to avoid symbols in which the time between adjacent pulses is zero, an additional guard chip may also be added to each symbol immediately following the pulse. DPIM may be expressed as

$$S_{DPIM}(t) = \begin{cases} P_c & nT_c \le t < (n+1)T_c \\ 0 & (n+1)T_c \le t < (n+k+1)T_c, \end{cases} \tag{5.49}$$

where T_c is the chip duration. Mapping between source and transmitted bits of 4-PPM and 4-DPIM is shown in Table 5.4, and this mapping is very well illustrated in Fig. 5.12 for the same transmitted source bits combination, i.e., 01 and 10.

It requires only chip synchronization and does not require symbol synchronization since each symbol is initiated with a pulse. It has higher transmission capacity as it eliminates unused time chips within each symbol. Assuming the symbol length

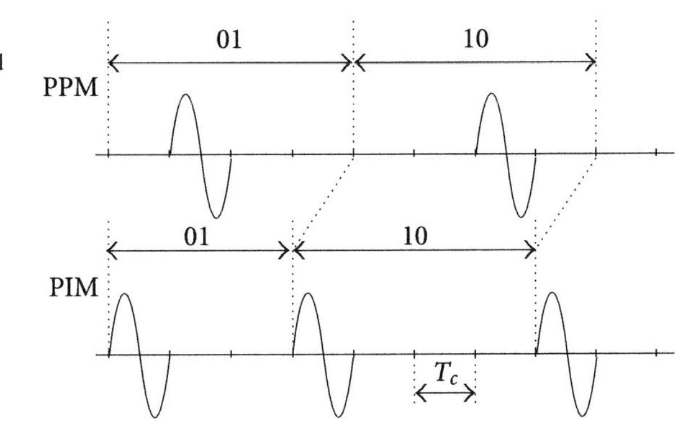

Fig. 5.12 Comparison of symbol structure for PPM and DPIM for same transmitted source bit combination, i.e., 01 and 10 [11]

Table 5.4 Mapping between 4-PPM and 4-DPIM chips

Source bits	4-PPM	4-PIM
00	1000	1(0)
01	0100	1(0)0
10	0010	1(0)00
11	0001	1(0)000

is random and uniformly distributed between 2 and $2^{\mathbb{M}}$ slots, the average bit rate R_b is given as

$$R_b = \frac{\mathbb{M}}{L_{avg} T_c}, \tag{5.50}$$

where $L_{avg} = \left(2^{\mathbb{M}} + 3\right)/2$ is the mean symbol length in the slot with zero guard chip. The bandwidth efficiency of DPIM is given as

$$\frac{R_b}{B} = \frac{2\mathbb{M}}{\left(2^{\mathbb{M}} + 3\right)}. \tag{5.51}$$

The packet error probability of DPIM is given as [10]

$$P_e = 0.5\text{erfc}\left[\frac{L_{avg} P_{avg} R_0 \sqrt{T_c}}{2\sqrt{2\sigma_n}}\right]. \tag{5.52}$$

Figure 5.13 shows the comparison of packet error rate performance of PPM and PIM for the same average power per symbol. It is observed from this figure that for same PER performance and packet length of 512 bits, 2-PIM has 4 dB lesser power requirement than 2-PPM. As the number of levels increases, PER performance degrades. Figure 5.14 gives the calculated packet error rate for DPIM, PPM, and OOK modulation schemes vs. average received irradiance based on a simple threshold detector. The background power is assumed to be $-10\,\text{dBm/cm}^2$ with

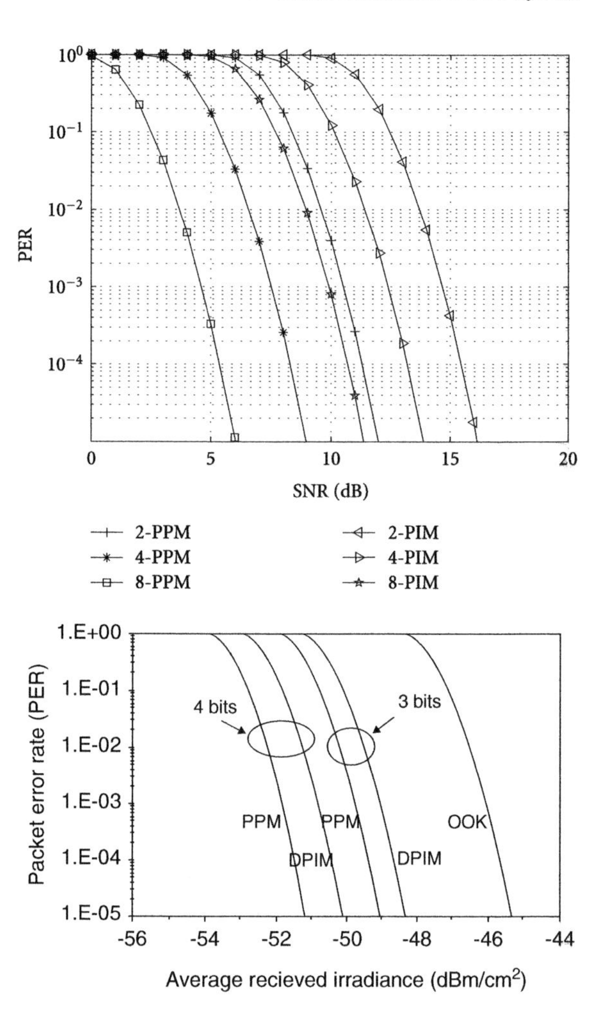

Fig. 5.13 Comparison of packet error rate performance of PPM and PIM schemes for modulation levels 2, 4, and 8 with same average power per symbol [11] (This image is reproduced here from the original article published under license to BioMed Central Ltd. This is an open access article distributed under the Creative Commons Attribution License, which permits unrestricted use, distribution, and reproduction in any medium, provided the original work is properly cited)

Fig. 5.14 Comparative packet error rate performance for DPIM, PPM, and OOK schemes vs. average received irradiance [10]

packet length of 1024 bits and data rate of 1 Mbps. It is observed from this figure that for the same packet error performance, 16-DPIM (no. of bits = 4) has about 5 dB power advantage over OOK, but requires approximately 1 dB more power than 16-PPM.

5.2.2.6 Dual Header-Pulse Interval Modulation

In dual header-pulse interval modulation (DHPIM), a symbol can have one of the two predefined headers depending on the input information as shown in Fig. 5.15.

The n_{th} symbol $S_n(h_n, d_n)$ of a DHPIM sequence is composed of a header h_n, which initiates the symbol, and information slots d_n. Depending on the most significant bit (MSB) of the input code word, two different headers are considered

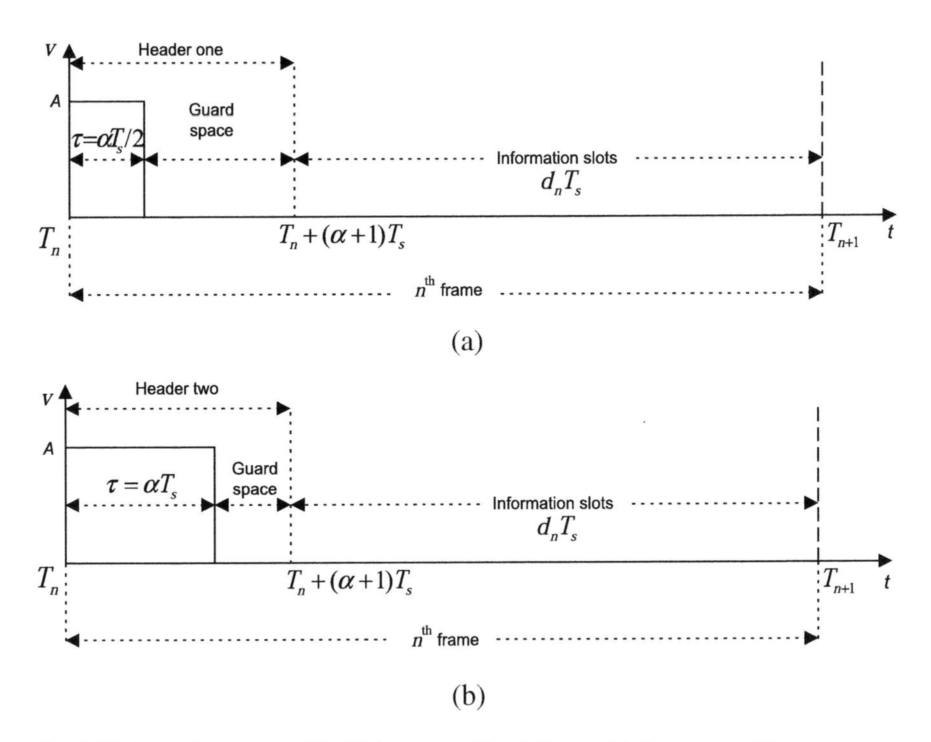

Fig. 5.15 Symbol structure of DHPIM scheme with (**a**) H_0 and (**b**) H_1 headers [12]

H_0 and H_1. If the MSB of the binary input word is equal to 0, then H_0 is used with d representing the decimal value of the input binary word. However, if MSB $= 1$ then H_1 is used with d being equal to the decimal value of the 1's complement of the input binary word. H_0 and H_1 have an equal duration of $T_h = (\alpha + 1)T_s$, where $\alpha > 0$ is an integer and T_s the slot duration, and are composed of a pulse and guard band. For H_0 and H_1, the pulse duration is $\alpha T_s/2$ and αT_s, respectively. The guard band duration depends on the header pulse duration and is $(\alpha + 2)T_s/2$ and T_s corresponding to H_0 and H_1, respectively. The information section is composed of d_n empty slots. The value of $d_n \in \{0,1,\ldots,2^{M-1} - 1\}$ is simply the decimal value of the remaining $M - 1$ bits of the M-bit input code word when the symbol starts with H_0 or the decimal value of its 1's complement when the symbol starts with H_1.

The header pulse has the dual role of symbol initiation and time reference for the preceding and succeeding symbols resulting in built-in symbol synchronization. Since the average symbol length can be reduced by a proper selection of α, DHPIM can offer shorter symbol lengths, improved transmission rate, and bandwidth requirements compared with the PPM, DPPM, and DPIM. Theoretically, it is possible to use a larger value of α; however this option increases the average symbol length unnecessarily, thus resulting in reduced data throughput. Therefore, $\alpha = 1$ or 2 is usually recommended.

Table 5.5 Mapping of 3-bit OOK words into PPM, DPPM, DHPIM, and DAPPM symbols

OOK	PPM ($\mathbb{M} = 8$)	DPPM ($\mathbb{M} = 8$)	DHPIM$_2$ ($\mathbb{M} = 8$)	DAPPM ($L_A = 2,\ \mathbb{M} = 4$)	DAPPM ($L_A = 4,\ \mathbb{M} = 2$)
000	10000000	1	100	1	1
001	01000000	01	1000	01	01
010	00100000	001	10000	001	2
011	00010000	0001	100000	0001	02
100	00001000	00001	110000	2	3
101	00000100	000001	11000	02	03
110	00000010	0000001	1100	002	4
111	00000001	00000001	110	0002	04

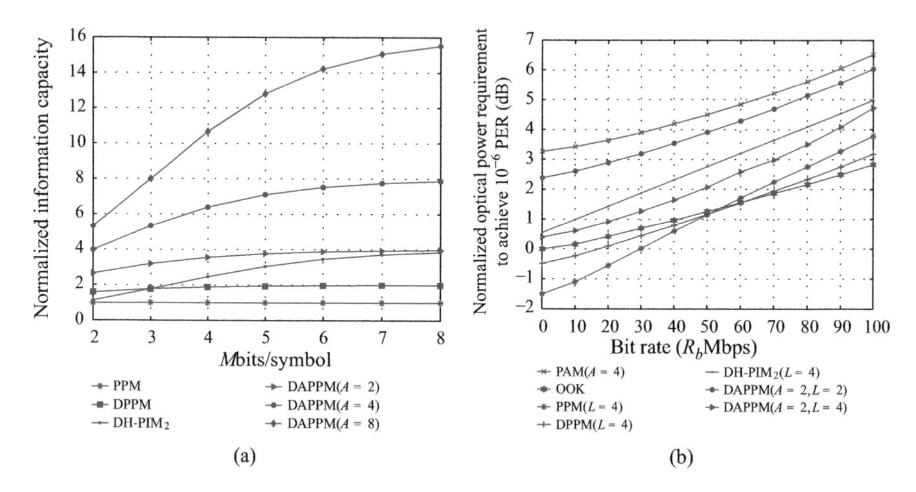

Fig. 5.16 Plot of variants of PPM for (**a**) capacity of variants of PPM normalized to capacity of OOK and (**b**) average optical power requirement to achieve packer error rate = 10^{-6} over dispersive channel [9]

The bandwidth efficiency of DHPIM is given as

$$\frac{R_b}{B} = \frac{2^{\mathbb{M}-1} + 2\alpha + 1}{2\mathbb{M}}. \tag{5.53}$$

The mapping of 3-bit OOK word for variants of PPM is shown in Table 5.5 for better understanding.

The plot of capacity and average optical power requirement of all the modulation schemes is shown in Fig. 5.16. It is observed from this figure that DAPPM provides higher transmission capacity than any other modulation scheme. The capacity of DH-PIM$_2$ is about the same as DAPPM ($A = 2$). Also, it has been observed that DAPPM is less power efficient and requires more average optical power than PPM and DPPM.

Table 5.6 Comparison of variants of PPM

Modulation schemes	$\mathbb{M}$−PPM	DPPM	DHPIM$_\alpha$	DAPPM	DPIM
Bandwidth (Hz)	$\frac{\mathbb{M}R_b}{\log_2\mathbb{M}}$	$\frac{(\mathbb{M}+1)R_b}{2\log_2\mathbb{M}}$	$\frac{(2^{\log_2\mathbb{M}-1}+2\alpha+1)R_b}{2\log_2\mathbb{M}}$	$\frac{(\mathbb{M}+A)R_b}{2\mathbb{M}A}$	$\frac{(\mathbb{M}+3)R_b}{2\log_2\mathbb{M}}$
PAPR	$\mathbb{M}$	$\frac{\mathbb{M}+1}{2}$	$\frac{2(2^{\log_2\mathbb{M}-1}+2\alpha+1)}{3\alpha}$	$\frac{A(\mathbb{M}+1)}{A+1}$	$\frac{\mathbb{M}+1}{2}$
Capacity	$\log_2\mathbb{M}$	$\frac{2\log_2\mathbb{M}}{\mathbb{M}+1}$	$\frac{2\mathbb{M}\log_2\mathbb{M}}{2(2^{\log_2\mathbb{M}-1}+2\alpha+1)}$	$\frac{2\mathbb{M}\log_2(\mathbb{M}\cdot A)}{\mathbb{M}+A}$	$\frac{2\log_2\mathbb{M}}{\mathbb{M}+3}$

5.3 Summary

In this chapter, both coherent and noncoherent modulation schemes have been discussed. The selection criteria of these modulation schemes will depend upon the application, whether it is power limited or bandwidth limited. Some modulation schemes are power efficient while others are bandwidth efficient. Most popularly used modulation scheme in FSO communication is OOK due to its simplicity. PPM is having low peak-to-average power ratio, hence making it a more power-efficient scheme. For this reason, it is considered for deep space laser communication. However, it suffers from poor bandwidth efficiency. In order to overcome this problem, variants of PPM are introduced that are more bandwidth efficient than PPM, but they result in an increase in system design complexity. A comparison of bandwidth requirement, peak-to-average power ratio (PAPR), and capacity for all the modulation schemes is given in Table 5.6.

It is seen that DAPPM is capable of providing better bandwidth efficiency; however owing to its longer symbol duration, DAPPM is more susceptible to intersymbol interference than any other variant of PPM.

Bibliography

1. A. Jurado-Navas, A. Garcia-Zambrana, A. Puerta-Notario, Efficient lognormal channel model for turbulent FSO communications. Electron. Lett. **43**(3), 178–179 (2007)
2. Q. Shi, Y. Karasawa, An accurate and efficient approximation to the Gaussian Q-function and its applications in performance analysis in Nakagami-m fading. IEEE Commun. Lett. **15**(5), 479–481 (2011)
3. Q. Liu, D.A. Pierce, A note on Gauss-Hermite quadrature. J. Biom. **81**(3), 624–629 (1994)
4. S. Haykin, M. Moher, *Communication Systems* (John Wiley & Sons, Inc., USA, 2009)
5. A. Kumar, V.K. Jain, Antenna aperture averaging with different modulation schemes for optical satellite communication links. J. Opt. Netw. **6**(12), 1323–1328 (2007)
6. Z. Ghassemlooy, W.O. Popoola, Terrestial free-space optical communications, in *Mobile and Wireless Communications Network Layer and Circuit Level Design*, ch. 17 (InTech, 2010), pp. 356–392. doi:10.5772/7698

7. K. Kiasaleh, Performance of APD-based, PPM free-space optical communication systems in atmospheric turbulence. IEEE Trans. Commun. **53**(9), 1455–1461 (2005)
8. D. Shiu, J.M. Kahn, Differential pulse-position modulation for power-efficient optical communication. IEEE Trans. Commun. **47**(8), 1201–1210 (1999)
9. U. Sethakaset, T.A. Gulliver, Differential amplitude pulse-position modulation for indoor wireless optical communications. J. Wirel. Commun. Netw. **1**, 3–11 (2005)
10. Z. Ghassemlooy, A.R. Hayes, N.L. Seed, E. Kaluarachehi, Digital pulse interval modulation for optical wireless communications. IEEE Commun. Mag. **98**, 95–99 (1998)
11. M. Herceg, T. Svedek, T. Matic, Pulse interval modulation for ultra-high speed IR-UWB communications systems. J. Adv. Signal Process. (2010). doi:10.1155/2010/658451
12. N.M. Aldibbiat, Z. Ghassemlooy, R. McLaughlin, Performance of dual header-pulse interval modulation (DH-PIM) for optical wireless communication systems. Proc. SPIE Opt. Wirel. Commun. III **4214**, 144–152 (2001)

Chapter 6
Link Performance Improvement Techniques

6.1 Aperture Averaging

In this case, the size of the receiver aperture is made larger than the operating wavelength so that the photon collecting capability of the detector increases. As a result, the constructive and destructive interferences average out to a larger extent and mitigate the effect of atmospheric turbulence. As per aperture averaging theory [1], the parameter that quantifies the reduction in fading is called aperture averaging factor, A_f given as

$$A_f = \frac{\sigma_I^2\,(D_R)}{\sigma_I^2\,(0)}, \tag{6.1}$$

where $\sigma_I^2\,(D_R)$ and $\sigma_I^2\,(0)$ denote the scintillation index for a receiver lens of diameter D_R and a "point receiver" ($D_R \approx 0$), respectively. The theory of aperture averaging has been extensively developed for plane and spherical waves in weak atmospheric turbulence conditions. For weak atmospheric turbulence, in the absence of inner and outer scale, the expressions for scintillation index for plane waves, $\sigma_{I,Pl}^2$, and spherical wave, $\sigma_{I,Sp}^2$, are given as [2]

$$\sigma_{I,Pl}^2 = \exp\left[\frac{0.49\sigma_R^2}{\left(1 + 0.65d^2 + 1.11\sigma_R^{12/5}\right)^{7/6}} + \frac{0.51\sigma_R^2\left(1 + 0.69\sigma_R^{12/5}\right)^{-5/6}}{1 + 0.90d^2 + 0.62d^2\sigma_R^{12/5}}\right] - 1 \tag{6.2}$$

H. Kaushal et al., *Free Space Optical Communication*, Optical Networks,
DOI 10.1007/978-81-322-3691-7_6

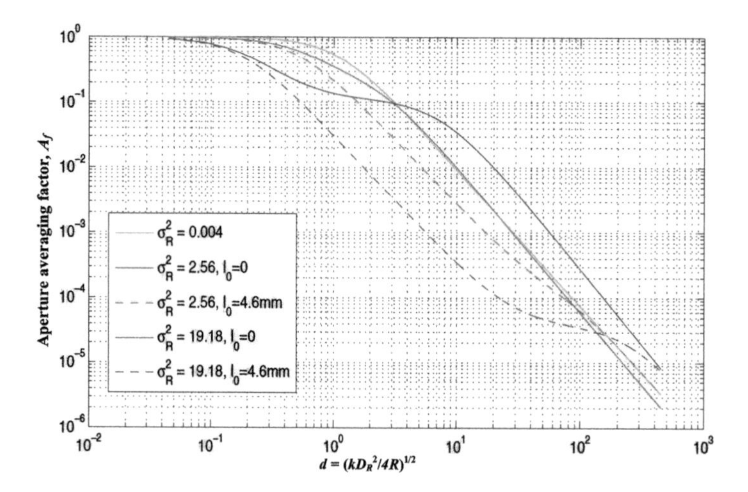

Fig. 6.1 Variation of aperture averaging factor, A_f with normalized receiver lens radius, d for various atmospheric turbulence conditions [3]

and

$$\sigma_{I,Sp}^2 = \exp\left[\frac{0.49\beta_o^2}{\left(1+0.18d^2+0.56\beta_o^{12/5}\right)^{7/6}} + \frac{0.51\beta_o^2\left(1+0.69\beta_o^{12/5}\right)^{-5/6}}{1+0.90d^2+0.62d^2\beta_o^{12/5}}\right] - 1, \text{ respectively.} \tag{6.3}$$

In the above equations, σ_R^2 is Rytov variance (discussed in Chap. 2), $d = \sqrt{\frac{kD_R^2}{4R}}$ is the circular aperture radius scaled by Fresnel length ($\mathcal{F} = \sqrt{R/k}$), k ($= 2\pi/\lambda$) is the wave number, R is the propagating path and D_R is the size of receiver aperture, and $\beta_o^2 = 0.5C_n^2k^{7/6}R^{11/6}$. It is observed from Fig. 6.1 that in case of weak to moderate atmospheric turbulence, aperture averaging factor, A_f decreases effectively for $D_R > \sqrt{\lambda R}$ as it averages out the irradiance fluctuations over turbulence. However, for strong atmospheric turbulence conditions, when $\sigma_R^2 \gg 1$, leveling effect is observed for $r_0 < D_R < R/kr_0$[4].

6.1.1 *Aperture Averaging Factor*

6.1.1.1 Plane Wave with Small l_o

This is the case when inner scale of turbulent eddies, l_o is much smaller than the Fresnel length $\left(=\sqrt{R/k}\right)$ or atmospheric coherence length, r_0, whichever is the smallest. Aperture averaging factor for plane wave propagating in weak atmospheric turbulence and following horizontal link has been approximated by Churnside [1, 5] and is given as

$$A_f = \left[1 + 1.07\left(\frac{kD_R^2}{4R}\right)^{7/6}\right]^{-1}. \tag{6.4}$$

Other approximations made by Andrew [6] showed slightly better results. Andrew approximated the aperture averaging factor as

$$A_f = \left[1 + 1.602\left(\frac{kD_R^2}{4R}\right)\right]^{-7/6}. \tag{6.5}$$

Equation (6.5) gives 7 % better results than Eq. (6.4) at $d = \sqrt{kD_R^2/4R} = 1$. It is clearly evident from (6.4) that for a given operating wavelength and a link length, as the aperture size increases, the impact of aperture averaging factor also increases, thus reducing the irradiance fluctuations or scintillation. In case of ground-to-satellite communication, i.e., vertical or slant path propagation, the aperture averaging factor is given by [7]

$$A_f = \frac{1}{1 + A_o^{-1}\left(\frac{D^2}{\lambda h_o sec\theta}\right)^{7/6}}, \tag{6.6}$$

where A_o is a constant approximately equal to 1.1, θ is the zenith angle, and h_o is given by

$$h_o = \left[\frac{\int C_n^2(h)\, h^2 dh}{\int C_n^2(h)\, h^{5/6} dh}\right]^{6/7}. \tag{6.7}$$

The decrease in aperture averaging factor, A_f, with the increase in aperture diameter, D_R, as shown in Fig. 6.2a, b, clearly indicates the reduction in scintillation. However, not much improvement in aperture averaging is observed with the change in zenith angle, θ as shown in Fig. 6.2b.

6.1.1.2 Plane Wave with Large l_o

This is the case when inner scale is much larger than Fresnel Length, i.e., $l_o \gg \sqrt{R/k}$. Aperture averaging factor for this case is approximated using Tatarskii or Hill spectrum [8] and is given as

$$A_f = \left[1 + 2.21\left(\frac{D_R}{l_0}\right)^{7/3}\right]^{-1}, \quad \text{for } D_R \gg 1. \tag{6.8}$$

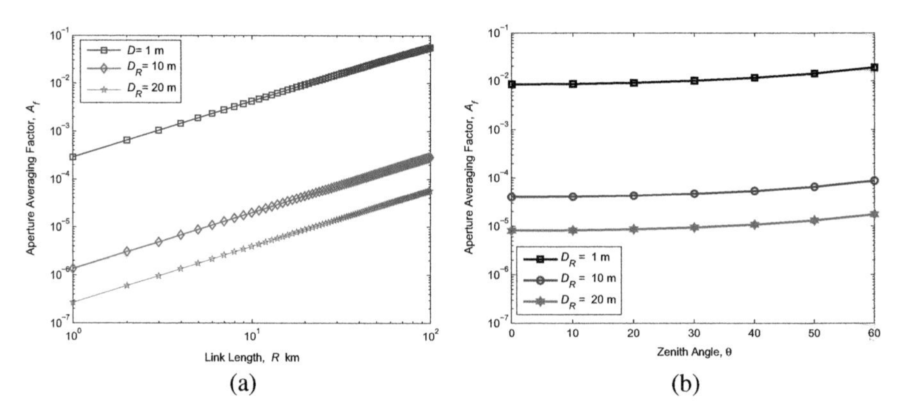

Fig. 6.2 Variation of aperture averaging factor, A_f for different aperture diameters, D_R with (**a**) horizontal link propagation and (**b**) slant link propagation

The improved version of Eq. (6.8) is [1]

$$A_f = \left[1 + 2.19\left(\frac{D_R}{l_0}\right)^2\right]^{-7/6}, \quad \text{for } 0 \leq \frac{D_R}{l_0} \leq 0.5. \tag{6.9}$$

Dependance of aperture averaging factor, A_f, on aperture diameter, D_R, is same as in the case of plane wave with small l_0, i.e., Eq. (6.4).

6.1.1.3 Spherical Wave with Small l_o

The aperture averaging factor in this case is evaluated using Kolmogorov spectrum and can be approximated as

$$A_f = \left[1 + 0.214\left(\frac{kD_R^2}{4R}\right)^{7/6}\right]^{-1}. \tag{6.10}$$

The amount of aperture averaging factor is 86 % less than the theoretical value when $d = \sqrt{kD_R^2/4R} = 1$.

6.1.1.4 Spherical Wave with Large l_o

The aperture averaging factor in this case is approximated by [1] and is given below

$$A_f = \left[1 + 0.109\left(\frac{D_R}{l_o}\right)^{7/3}\right]^{-1}. \tag{6.11}$$

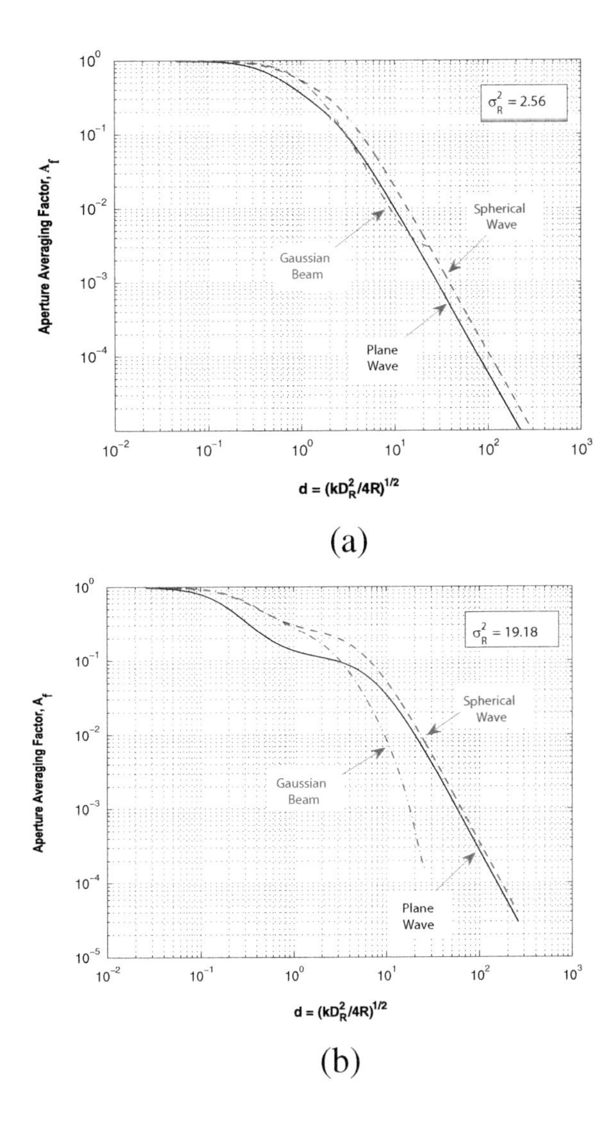

Fig. 6.3 Aperture averaging factor, A_f for different propagation models (i.e., plane, spherical and Gaussian) in (**a**) moderate and (**b**) strong atmospheric turbulence [9]

For the cases when l_o is 1.5 times the Fresnel length, it is recommended [1] to use small inner scale approximation when $l_o < 1.5\sqrt{R/k}$ and large inner scale approximation when $l_o \geq 1.5\sqrt{R/k}$. Andrew [6] suggested using small inner scale approximation when $kD_R^2/4R < 1$. Figure 6.3 shows the effect of aperture averaging for various propagation models in moderate and strong atmospheric turbulence.

6.2 Aperture Averaging Experiment

In order to verify the theory of aperture averaging, lots of experimentations have been carried out by various researchers. One such experiment as reported in [10] has been discussed here. The experimental results may tend to deviate from theoretical results owing to smaller propagation distance which changes the effective value of C_n^2. During experimentation, to see the effect of increase in receiver aperture diameter in atmospheric turbulent environment, an artificial turbulence is generated inside laboratory environment using optical turbulence generator (OTG) chamber as shown in Fig. 6.4a. The turbulence is generated inside the chamber by forced mixing of cold and hot air [10]. To build up varying strength of turbulence inside the OTG chamber, the air intake at one end of the chamber is kept at room temperature and the other side hot air is blown in using electrical heater. The temperature inside the chamber can be controlled by varying the variac connected to the heater, and thus varying strengths of C_n^2 can be generated. The velocity inflow of the air can be controlled by rpm of the fans connected to the OTG chamber. Therefore, by varying velocity and temperature, different strength of turbulence can be generated inside the chamber. The other parameters that were set for the laboratory experimentation are given in Table 6.1.

The laser beam is allowed to propagate through the OTG chamber perpendicular to the direction of the turbulence air flow as shown in Fig. 6.4b, and the abbreviated wave front is captured by the beam profiler placed at the other end. The whole setup was fixed on a vibration-free table, and the observations were taken in dark room environment. The first step in performing the experiment is to characterize the turbulence inside the OTG chamber. This is done by evaluating change in the instantaneous point of maximum irradiance due to change in the direction of the beam. In other words, as the beam propagates inside the OTG chamber, variance of beam wander has to be determined. The beam wander variance will then help in determining the value of C_n^2, the parameter that characterizes turbulence inside the

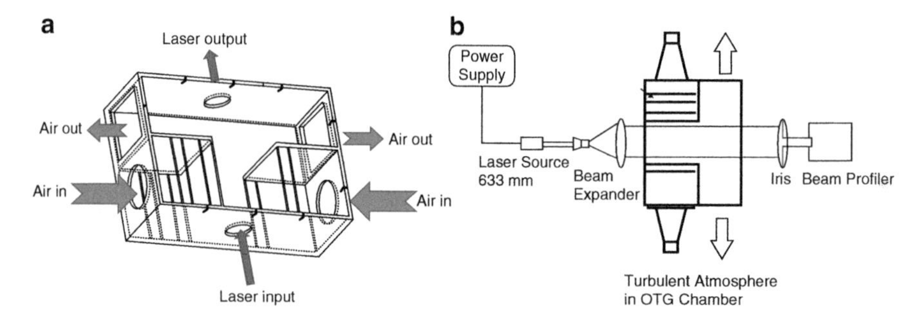

Fig. 6.4 Aperture averaging experiment. (**a**) Three-dimensional view of OTG chamber and (**b**) experimental setup

Table 6.1 Parameters used in laboratory experimentation

S.No	Parameters	Values
1.	Laser power	10 mW
2.	Tx. beam size (W_0)	0.48 mm
3.	Operating wavelength	633 and 542 nm
4.	OTG chamber dimensions	$20 \times 20 \times 20$ cm
5.	Receiver camera type	CCD type
7.	Exposure time	0.4 ms
6.	Camera resolution	1600×1200 pixels
7.	Pixel size	$4.5 \times 4.5\,\mu m$
8.	Temperature difference (δT)	20–100 K
9.	Zenith angle (θ)	$0°$
10.	Propagation length (R)	50 cm

Table 6.2 $C_n^2 \cdot \Delta R$ values for different temperature difference

S.No.	Temp diff (in K)	$C_n^2 \cdot \Delta R$ ($m^{-1/3}$)
1.	10	5.5×10^{-12}
2.	20	1.86×10^{-11}
3.	30	3.65×10^{-11}

OTG chamber. The beam wander variance for a collimated beam having beam size W_0 propagating at zero zenith angle is given by (refer Eq. (2.33))

$$\langle r_c^2 \rangle = 0.54R^2 \left(\frac{\lambda}{2W_0}\right)^2 \left(\frac{2W_0}{r_0}\right)^{5/3}. \tag{6.12}$$

Here, $\langle\rangle$ denotes the ensemble average and r_0 the atmospheric coherence length (also known as Fried parameter). Table 6.2 shows the experimental value of $C_n^2 \cdot \Delta R$ for different temperature inside the OTG chamber [11].

The experimental setup for determining the aperture averaging factor is shown in Fig. 6.4b. Beam expander is used to expand the transmitted beam width from the laser source. This expanded beam is transmitted through OTG chamber. At the receiver end, iris diaphragm is used to vary the receiver aperture. The beam from the iris is captured on to the beam profiler which is used to compute the beam variance. Figure 6.5 shows the analytical and experimental results, respectively, for aperture averaging factor vs. aperture diameter for $\sigma_I^2 = 0.03$. It is seen that in both the cases, the aperture averaging factor decreases with the increase in the aperture diameter. Further, a very marginal improvement in aperture averaging factor is seen after 4 mm aperture diameter for a given set of parameters. Although the trend of graph is same in both theoretical and experimental results, there seems to be deviation in the value of aperture averaging factor for a given aperture diameter. It is because the atmospheric turbulence is generated artificially inside the OTG chamber by varying the temperature and velocity of air flow over a limited distance. This will decrease the precision of C_n^2 value obtained with theoretical data.

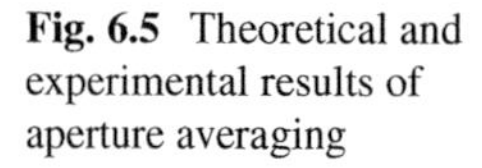

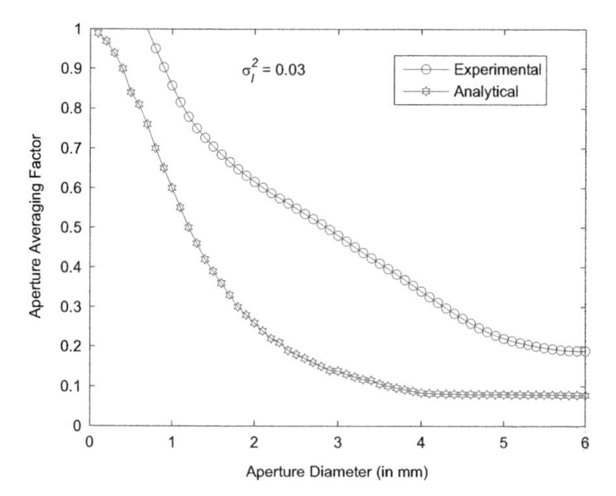

Fig. 6.5 Theoretical and experimental results of aperture averaging

6.3 Diversity

The performance of FSO communication system in turbulent atmospheric channel can be improved by employing diversity techniques. Diversity in wireless communication system can be of various types: time diversity, frequency diversity, or spatial diversity. In case of time diversity, multiple copies of same signal are transmitted using different time slots. In case of frequency diversity, the information is transmitted using different subcarriers to achieve good system performance maintaining sufficient separation between the carriers. In space diversity, multiple transmitters or receivers or both are used ensuring zero correlation between different antennae. This method of diversity gives maximum diversity gain provided the signals are received independently (zero correlation between signals). The concept of spatial diversity is to achieve the diversity gain of large aperture by using array of small detector apertures that are sufficiently separated in space. The required antenna separation depends upon beam divergence, operating wavelength, and atmospheric turbulence level.

In order to implement diversity with good diversity gain, estimate of the channel parameters has to be carried out accurately. The time-varying turbulent atmospheric channel is very difficult to understand. One approximate channel model is *wide sense stationary uncorrelated scattering channel.* In this channel model, the scatterers by different objects are assumed to be independent. The parameters that characterize such channel are as follows:

(i) Multipath spread, T_m: It tells us about the maximum delay between two propagation paths of significant power in the channel.
(ii) Coherence bandwidth, Δf_c: It is the maximum frequency interval over which two frequencies of signal are likely to experience comparable correlated degree of fading or highly correlated.

(iii) Coherence time, $\triangle t_c$: It is the maximum time interval over which the channel impulse response is essentially invariant or correlated.
(iv) Doppler spread, B_d: It gives the maximum range of Doppler shift.

Based on these channel parameters and the characteristics of the transmitted signal, time-varying fading channel can be characterized as follows:

(i) Frequency nonselective vs. frequency selective: If the transmitted signal bandwidth is small compared with $\triangle f_c$, then all frequency components of the signal would roughly undergo the same degree of fading. This type of channel is then classified as frequency nonselective (also called flat fading) channel. Since $\triangle f_c$ and $\triangle t_c$ have reciprocal relationship, the frequency nonselective channels have, therefore, large symbol duration (reciprocal of bandwidth) as compared with $\triangle t_c$. In this type of channel, the delay between different paths is relatively small with respect to the symbol duration. The gain and phase of the received signal is determined by superposition of all the copies of the signal that arrive within time interval of $\triangle t_c$.

On the other hand, if the bandwidth of the transmitted signal is large compared with $\triangle f_c$, then different frequency components of the signal would undergo different degrees of fading. This type of channel is then classified as frequency selective channel. Due to the reciprocal relationship, symbol duration of the transmitted signal is small compared with $\triangle t_c$. In this type of channel, delays between different paths can be relatively large with respect to the symbol duration, and, therefore, multiple copies of the signal could be received at the other end.
(ii) Slow vs. fast fading: If the symbol duration is small compared with $\triangle t_c$, then the channel is classified as slow fading. In this case, the channel is often modeled as time invariant over a number of symbol intervals. On the other hand, if the symbol duration is large than $\triangle t_c$, then the channel is considered as fast fading channel (also known as time selective fading). In this case, channel parameters cannot be estimated very easily.

Diversity technique is an attractive approach in atmospheric turbulent environment due to their inherent redundancy. This technique also significantly reduces the chances for temporary blockage of the laser beam by any kind of obstructions (e.g., birds). For an FSO system employing M transmit and N receive antennae, the received power at the nth receiver is given by

$$r_n = s\eta \sum_{m=1}^{M} I_{mn} + z(t), \tag{6.13}$$

where $s \epsilon \{0, 1\}$ is the transmitted bit information, η the optical to electrical conversion coefficient, and $z(t)$ the additive white Gaussian noise with zero mean and variance $\sigma_n^2 = N_0/2$. For a constant transmitted total power P, each beam will carry $1/M$th of the power P and therefore limiting the transmit power density (expressed as milliwatt per square centimeter). The number of multiple beams needed to achieve a given BER depends upon the strength of the atmospheric turbulence. However, the multiple beams cannot be increased indefinitely due to practical considerations such as available space, cost, laser efficiency in dividing the power, etc.

6.3.1 Types of Diversity Techniques

Diversity techniques are effective when different branches are independently faded or uncorrelated. It can be obtained in various ways as discussed below:

(i) Frequency or wavelength diversity: Here, the same information is transmitted on different carriers provided the frequency separation between the carriers should be at least equal to Δf_c. This will ensure that different branches undergo different fading.

(ii) Time diversity: Time diversity is achieved by transmitting the same bit of information repetitively at short intervals of time. The interval between symbols should be at least equal to Δt_c so that signals in different branches undergo different degree of fading. The time interval depends on the fading rate, and it increases with the decrease in the rate of fading.

(iii) Space diversity: Space diversity also called antenna diversity can be achieved by placing multiple antennae at the transmitter and/or at the receiver. The spacing between the antennae should be more than the coherence length, r_0 of the atmosphere so that signals in different branches undergo different degree of fading and are independent. As operating wavelength is very small for optical frequencies, therefore, very small spacing between the antennae will suffix the requirement to achieve sufficient diversity gain (Figs. 6.6 and 6.7).

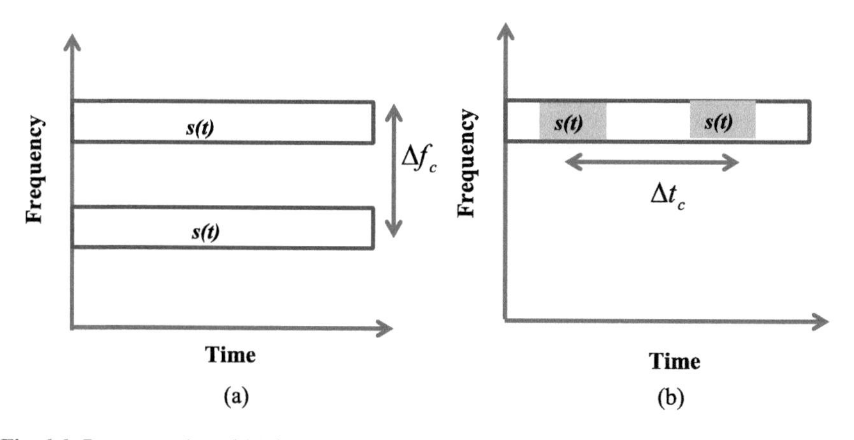

Fig. 6.6 Representation of (**a**) frequency diversity and (**b**) time diversity

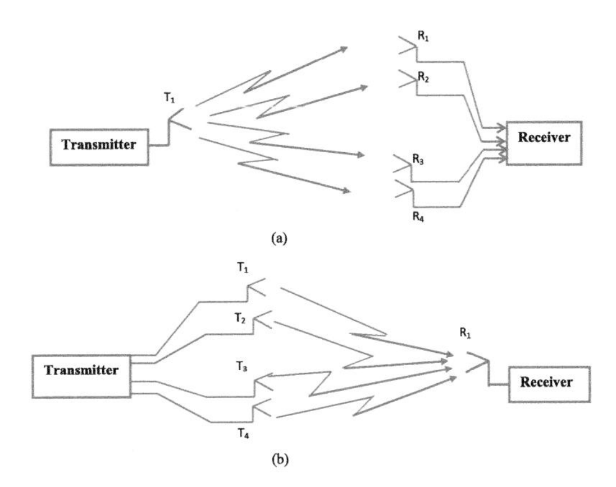

Fig. 6.7 Representation of (**a**) receive and (**b**) transmit spatial diversity

Out of all these diversity techniques, space diversity or combination of space and time diversity is most commonly used in FSO communication.

6.3.2 Diversity Combining Techniques

The idea of diversity is to combine several copies of the transmitted signal, which have undergone independent fading in order to achieve the diversity gain. There are many diversity combining techniques that help in improving the received SNR. For slowly varying flat fading channels, the received signal at ith branch can be expressed as

$$r_i(t) = \alpha e^{j\theta_i} s(t) + z_i(t) \text{ , for } i = 1, 2, \ldots\ldots .M, \tag{6.14}$$

where $s(t)$ is the transmitted signal, $z_i(t)$ additive white Gaussian noise, and $\alpha e^{j\theta_i}$ the fading coefficient of each ith branch. For M independent branches, M replicas of the transmitted signal are obtained as

$$r = \left[r_1(t)\ r_2(t) \cdots r_M(t) \right]. \tag{6.15}$$

This M diverse signal will help in achieving diversity gain up to M by averaging over multiple independent signal paths and hence improve the performance of the system in turbulent atmosphere. It has been observed that diversity gain increases with the increase in the turbulence of the atmosphere and number of transmit/receive antennae. There are various combining techniques that will combine the received signal in an optimum or near optimum way in order to achieve good diversity gain at the receiver. Some of the most commonly used combining techniques are discussed below:

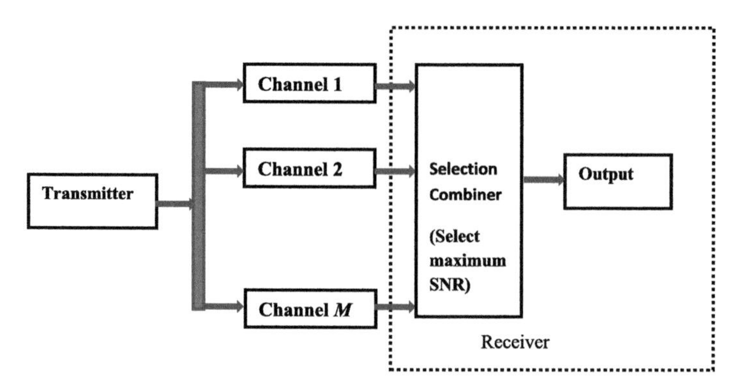

Fig. 6.8 Selection combining

(i) Selection combining (SC): This combining technique selects the branch with the strongest SNR as shown in Fig. 6.8.

The larger the number of available branches, the higher the probability of having a larger SNR at the receiver. For a flat fading channel, the instantaneous SNR in the ith branch is given by

$$\gamma_i = \frac{E}{N_0} |h_i|^2, \tag{6.16}$$

where h_i is the channel (complex) gain, E the symbol energy, N_0 the noise spectral power density assumed to be same in all the branches. The instantaneous SNR at selection combiner output is given by

$$\gamma_{sc} = \max \{\gamma_1, \gamma_2, \ldots.\gamma_M\}. \tag{6.17}$$

The pdf of instantaneous SNR of all the branches follows exponential distribution as given below:

$$f(\gamma_i) = \frac{1}{\gamma_{av}} \exp^{-\frac{\gamma_i}{\gamma_{av}}}, \text{ for } \gamma_i > 0, \tag{6.18}$$

where $\gamma_{av} = \overline{\gamma_i} = \frac{E}{N_0} \left|\overline{h_i}\right|^2$ is the average SNR. The associated cumulative distribution function (CDF) is given by

$$P(\gamma_i \le \gamma) = \int_{-\infty}^{\gamma} f(\gamma_i)\, d\gamma_i = 1 - \exp^{-\gamma/\gamma_{av}}. \tag{6.19}$$

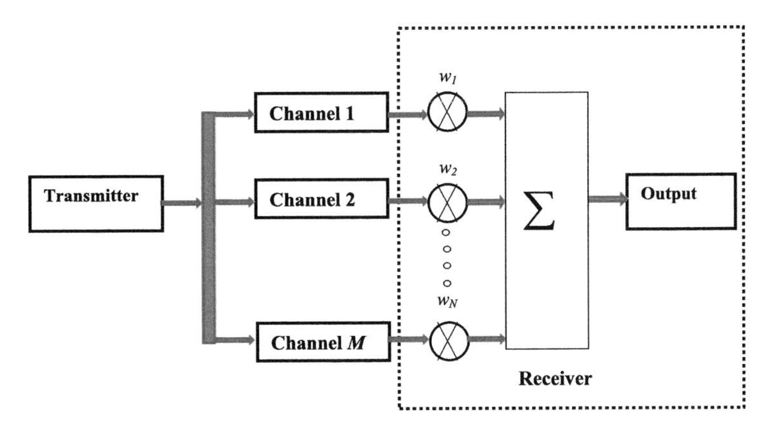

Fig. 6.9 Maximum ratio combining

The CDF for instantaneous SNR after selection combining can be expressed as:

$$\begin{aligned} F(\gamma_{sc}) = & \quad P(\gamma_i \leq \gamma_{sc},\ i = 1, 2\ldots M) = \prod_{i=1}^{M} P(\gamma_i \leq \gamma_{sc}) \\ & = \prod_{i=1}^{M} \left(1 - \exp^{-\gamma/\gamma_{av}}\right) = \left(1 - \exp^{-\gamma/\gamma_{av}}\right)^M, \gamma_{sc} \geq 0. \end{aligned} \tag{6.20}$$

This is also the outage probability (P_{out}) of ith branch which is another performance criterion characteristic of diversity systems operating over fading channels. Outage probability is defined as the probability of the instantaneous error probability exceeding a specified value or equivalently the probability that the output SNR falls below a certain specified threshold, γ_{th}. This is basically CDF of γ, evaluated at $\gamma = \gamma_{th}$.

(ii) Maximum ratio combining (MRC): In this combining technique, different weights are assigned to diversity branches for optimizing SNR as shown in Fig. 6.9. In this case, the baseband signal of different diversity branches can be expressed as

$$r_i(t) = h_i s(t) + z_i(t), \tag{6.21}$$

where h_i is the complex channel gain, $s(t)$ the transmitted signal, and $z_i(t)$ additive white Gaussian noise. The received signal is fed to the linear combiner whose output can be expressed as

$$\begin{aligned} y(t) &= \sum_{i=1}^{M} w_i r_i \\ &= \sum_{i=1}^{M} w_i \left(h_i s(t) + z_i(t)\right) \end{aligned}$$

$$= s(t) \cdot \sum_{i=1}^{M} w_i h_i + \sum_{i=1}^{M} w_i z_i(t). \tag{6.22}$$

The instantaneous signal and noise power at the combiner output is given as

$$\sigma_y^2 = E\left[|s(t)|^2\right] \cdot \left|\sum_{i=1}^{M} w_i h_i\right|^2 \tag{6.23}$$

and

$$\sigma_{nc}^2 = \sigma_n^2 \cdot \sum_{i=1}^{M} |w_i|^2, \text{ respectively.} \tag{6.24}$$

The instantaneous SNR at the combiner output is given as

$$\gamma_{MRC} = \gamma_{av} \frac{\left|\sum_{i=1}^{M} w_i h_i\right|^2}{\sum_{i=1}^{M} |w_i|^2}. \tag{6.25}$$

According to the Schwarz inequality for complex parameters,

$$\left|\sum_{i=1}^{M} w_i h_i\right|^2 \leq \sum_{i=1}^{M} |w_i|^2 \cdot \sum_{i=1}^{M} |h_i|^2, \tag{6.26}$$

which holds true when $w_i = a h_i^*$, where a is an arbitrary constant. If $a = 1$, the instantaneous SNR is maximized when weights are chosen to be conjugate of channel gain, i.e., $w_i = h_i^*$, and it becomes

$$\gamma_{MRC} = \gamma_{av} \sum_{i=1}^{M} |h_i|^2 = \sum_{i=1}^{M} \gamma_i, \tag{6.27}$$

where γ_i is the instantaneous SNR of the ith branch. The pdf of the SNR at the combiner output is given as

$$f(\gamma_{MRC}) = \frac{1}{(M-1)!} \cdot \frac{\gamma_{MRC}^{M-1}}{\gamma_{av}^{M}} \exp^{-\gamma_{MRC}/\gamma_{av}}. \tag{6.28}$$

The outage probability of MRC can then be obtained from CDF as follows:

$$P(\gamma_{MRC} \leq \gamma) = \int_{-\infty}^{\gamma} f(\gamma_{MRC})\, d\gamma_{MRC}$$

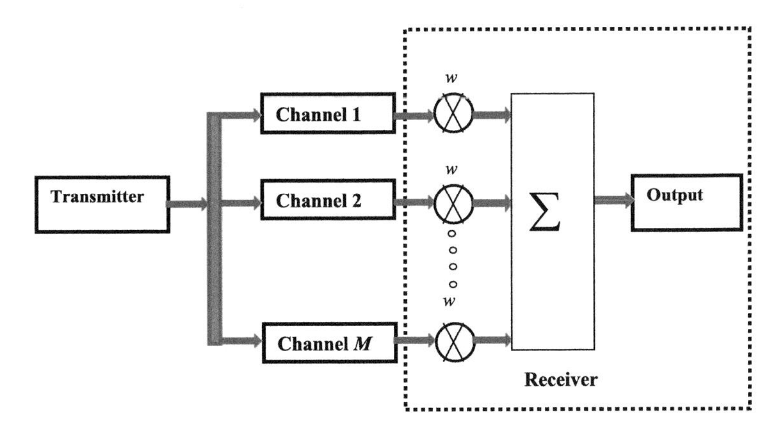

Fig. 6.10 Equal gain combining

$$= 1 - \int_{\gamma}^{\infty} f(\gamma_{MRC})\, d\gamma_{MRC}$$

$$= 1 - \exp^{-\gamma_{MRC}/\gamma_{av}} \sum_{i=1}^{M} \frac{\left(\gamma_{MRC}/\gamma_{av}\right)^{i-1}}{(i-1)!}. \tag{6.29}$$

(iii) Equal gain combining (EGC): In this combining technique, equal weights are assigned to every branch and this reduces the complexity of the receiver as shown in Fig. 6.10. Comparing the performance of all the diversity combining techniques, it is observed that the performance of MRC outperforms all other combining techniques although it is difficult to implement. SC is the simplest combining technique; however, its performance is not up to the mark. EGC performs very close to MRC with lesser complexity. If the individual branch is not independent, then the performance of EGC technique degrades.

6.3.3 Alamouti's Transmit Diversity Scheme

In case of transmit diversity, there are multiple transmit antennae and one receiving antenna. It is also termed as multiple input single output (MISO). It can be achieved by transmitting the same symbol over different antennae during different time intervals. However, this is not very useful technique as it is repeating the same symbol in space and causes processing delays. The other way is to transmit different time diversity codes over different antennae at the same time. Still, there were some problems with this technique when multiple signal may produce nulls in the radiation at certain angles. Another technique is proposed by Alamouti that utilizes both time and space diversity, and it is also known as space time coding [12]. This transmit diversity technique makes use of two transmit antennae and one receive antenna to provide the same diversity as in MRC receiver using one transmit and two receive antennae. This scheme can be generalized to two transmit antenna and M receive antennae to provide diversity order of $2M$. It improves the error performance, data rate, or capacity of wireless communication system. The salient features associated with this diversity scheme are described below:

(i) In this scheme, the redundancy is applied in space across multiple antennae, not in time or frequency. Therefore, this scheme does not require any bandwidth expansion.
(ii) No feedback is required from receiver to transmitter and, therefore, it has less computational complexity.
(iii) This scheme is the cost-effective way to improve the system capacity or increase the coverage range in fading environment without a complete redesign of the existing system.

6.3.4 *Two Transmitter and One Receiver Scheme*

The block diagram of Alamouti's diversity is shown in Fig. 6.11. The scheme uses two transmit antennae and one receive antenna. At the transmitter, the information symbols will be encoded specially for transmit diversity system and transmitted toward the receiver. At the receiver, the received symbols are combined using combiner, and decision is made based on maximum likelihood detection.

Let $[s_0, s_1]$ be the complex data to be transmitted using Alamouti's space time encoding scheme as shown in Table 6.3.

At antennae 0

$A = \left[s_0, -s_1^*\right]$

At antennae 1

$B = \left[s_1, s_0^*\right]$

At a given symbol period, two symbols are simultaneously transmitted from the two antennae. During first symbol period, signal s_0 is transmitted from antenna 0 and signal s_1 is transmitted from antenna 1. During the next symbol period, $-s_1^*$ is transmitted from antenna 0, and signal s_0^* is transmitted from antenna 1 where "*" implies complex conjugate operation. This type of encoding is done in space and time and, therefore, termed as space time coding. However, if instead of two symbol period, two adjacent carriers are used, then the coding scheme is termed as space frequency coding. In this diversity scheme, the property of orthogonality is used, i.e., $A.B^T = 0$ as there is no co-phasing between the channels.

Let $h_0(t)$ and $h_1(t)$ be the complex channel gain between the receiver and transmit antennae 0 and 1, respectively. Assuming constant fading between two consecutive symbols, the channel state information can be expressed as

$$h_0(t) = h_0 = \alpha_0 \exp^{j\theta_0} \tag{6.30}$$

$$h_1(t) = h_1 = \alpha_1 \exp^{j\theta_1} \tag{6.31}$$

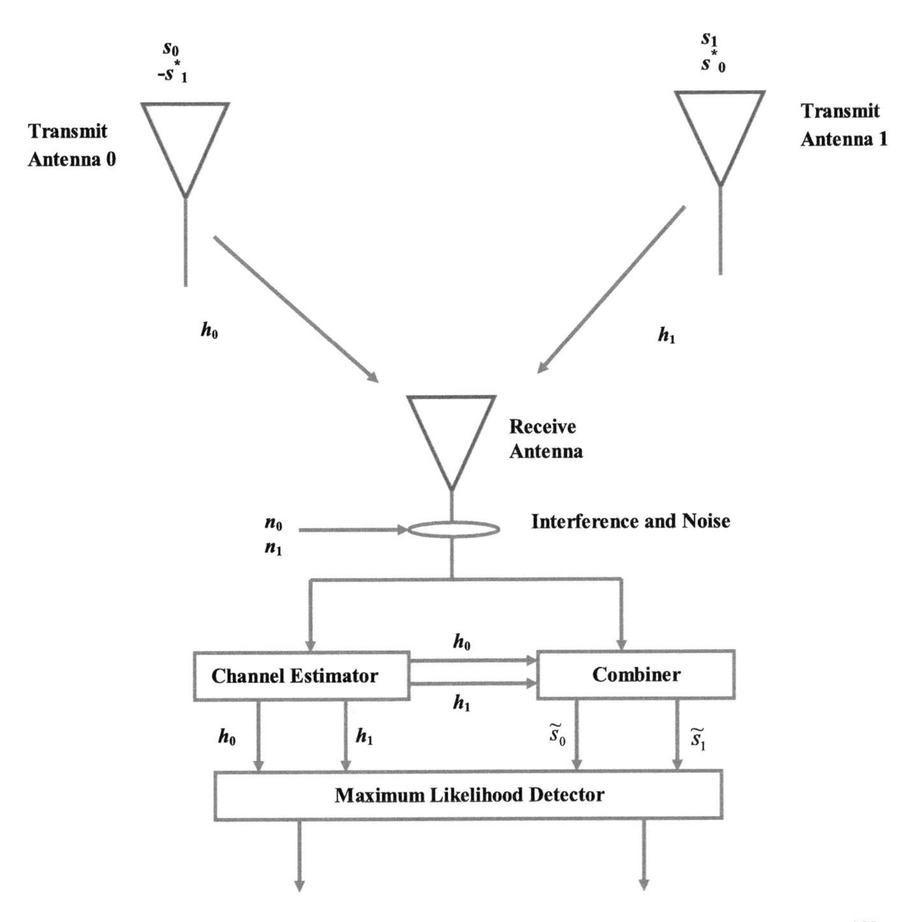

Fig. 6.11 Alamouti's transmit diversity scheme with two transmit and one receive antennae [12]

Table 6.3 Alamouti's space time encoding scheme for two-branch transmit diversity scheme

Symbol period	Antenna 0	Antenna 1
Time t	s_0	s_1
Time $t + T$	$-s_1^*$	s_0^*

The received signals can then be expressed as

$$r_0 = r(t) = h_0 s_0 + h_1 s_1 + n_0 \tag{6.32}$$

$$r_1 = r(t + T) = -h_0 s_1^* + h_1 s_0^* + n_1 \tag{6.33}$$

where T is the symbol duration, n_0 and n_1 are the complex random variables representing noise and interference, respectively. These received signals are then fed to the combiner that generates two signals as given below

$$\widetilde{S_0} = h_0^* r_0 + h_1 r_1^* \tag{6.34}$$

$$\widetilde{S_1} = h_1^* r_0 - h_0 r_1^* \tag{6.35}$$

Substituting Eqs. (6.30), (6.31), (6.32), and (6.33) in Eqs. (6.34) and (6.35), we get

$$\widetilde{S_0} = \left(\alpha_0^2 + \alpha_1^2\right) s_0 + h_0^* n_0 + h_1 n_1^* \tag{6.36}$$

$$\widetilde{S_1} = \left(\alpha_0^2 + \alpha_1^2\right) s_1 - h_0 n_1^* + h_1^* n_0 \tag{6.37}$$

The signals are then fed to maximum likelihood detector for decision purpose. These combined signals are equivalent to that obtained from one transmitter and two receiver of MRC receiver. Therefore, the resulting diversity order of two-branch transmit diversity scheme is same as two-branch MRC receiver.

6.3.5 BER Performance with and Without Spatial Diversity

For multiple transmitted beams in the presence of turbulence, the conditional BER is obtained by averaging over irradiance fluctuations statistics and is given as

$$P_e = \int_0^\infty f_I(I)\, Q\left(\sqrt{SNR(I)}\right) dI \tag{6.38}$$

BER expression for lognormal modeled FSO system can be written as

$$P_e = \int_0^\infty \frac{1}{I\sqrt{2\pi\sigma_l^2}} \exp\left\{-\frac{\left(\ln(I/I_0) + \sigma_l^2/2\right)^2}{2\sigma_l^2}\right\} Q\left(\sqrt{SNR(I)}\right) dI \tag{6.39}$$

The above expression can be solved using an alternative representation of Q-function [13, 14] together with the Gauss-Hermite quadrature integration. The equation can be rewritten as

$$P_e \cong \int_{-\infty}^\infty \frac{1}{\sqrt{\pi}} \exp\left(-x^2\right) Q\left(\frac{R_0 A I_0 \exp\left(\sqrt{2}\sigma_l x - \sigma_l^2/2\right)}{\sqrt{2}\sigma_n}\right) dx \tag{6.40}$$

where $x = \left(\frac{\ln(I/I_0)+\sigma_l^2/2}{\sqrt{2}\sigma_l}\right)$. The above equation can be approximated using Gauss-Hermite quadrature integration as given below:

$$\int_{-\infty}^\infty f(x) \exp\left(-x^2\right) dx \cong \sum_{i=1}^m w_i f(x_i) \tag{6.41}$$

where x_i and w_i represent the zeros and weights of the mth-order Hermite polynomials.

Using Eq. (6.40), the BER expression for coherent BPSK can be written as

$$P_e\,(\text{SC-BPSK}) \cong \frac{1}{\sqrt{\pi}} \sum_{i=1}^{m} w_i Q\left(K \exp\left[\left(\sqrt{2}\sigma_l x_i - \sigma_l^2/2\right)\right]\right) \quad (6.42)$$

where $K = R_0 A I_0 / \sqrt{2}\sigma_n$. Similarly, the BER for SC-QPSK is given by

$$P_e\,(\text{SC-QPSK}) \cong \frac{1}{\sqrt{\pi}} \sum_{i=1}^{m} w_i Q\left(\frac{K \exp\left[\left(\sqrt{2}\sigma_l x_i - \sigma_l^2/2\right)\right]}{\sqrt{2}}\right) \quad (6.43)$$

For transmit diversity, the effective variance in Eqs. (6.42) and (6.43) is scaled by the number of transmit antennae, i.e., $\widehat{\sigma_l^2} = \sigma_l^2/M$. It may be mentioned that this linear scaling sometimes does not hold if there is a correlation among the transmitting antennae as it reduces the achievable diversity gain. In that case, the effective variance is given by

$$\widehat{\sigma_l^2} = \frac{\sigma_l^2}{M} + \frac{1}{M^2} \sum_{\substack{p=1,\\ p \neq l}}^{M} \Gamma_{pl} \quad (6.44)$$

where Γ_{pl} $(p, l = 1, 2, \ldots M,\ p \neq l)$ are the correlation coefficients [15]. The performance of FSO link for SC-BPSK and SC-QPSK modulation schemes is computed using Eqs. (6.42), (6.43), and (6.44) for different values of correlation among the transmitting beams, i.e., for $\rho = 0.0, 0.3$, and 0.7. Figure 6.12a, b shows the computed results for SC-BPSK and SC-QPSK modulation schemes, respectively, without any correlation among the transmitting beams, i.e., $\rho = 0.0$. In presence of correlation, computed results for $\rho = 0.3$ and 0.7 are shown in Fig. 6.13a, b for SC-BPSK and in Fig. 6.14a, b for SC-QPSK modulation schemes, respectively.

The following conclusions are made from Figs. 6.12, 6.13, and 6.14:

(i) The degradation in the performance due to atmospheric turbulence is more in SC-QPSK than in SC-BPSK.
(ii) In both the cases, there is an improvement in the performance as the number of transmitting antennae is increased. However, the level of improvement is more in SC-BPSK as compared to SC-QPSK.
(iii) The improvement with spatial diversity is significant when the turbulence level is relatively high. This is true for both SC-BPSK and SC-QPSK modulation schemes.
(iv) The results at (ii) and (iii) are valid irrespective of the correlation coefficient among the transmit antenna beams.

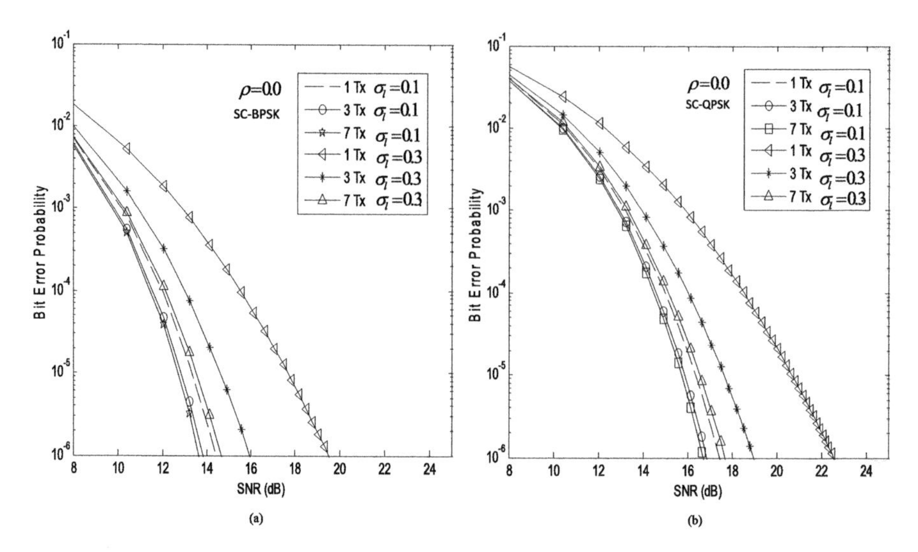

Fig. 6.12 Bit error probability vs. SNR with spatial diversity in weak turbulence (σ_l = 0.1 and 0.3) for subcarrier (**a**) BPSK and (**b**) QPSK modulation schemes when there is no correlation among transmitted antenna beams ($\rho = 0.0$)

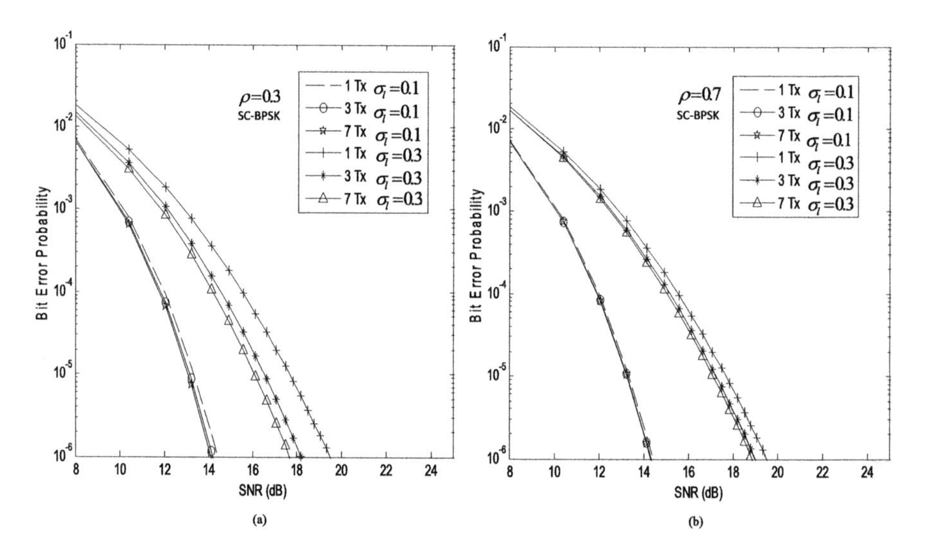

Fig. 6.13 Bit error probability vs. SNR with spatial diversity in weak turbulence (σ_l = 0.1 and 0.3) for SC-BPSK (**a**) $\rho = 0.3$ and (**b**) $\rho = 0.7$

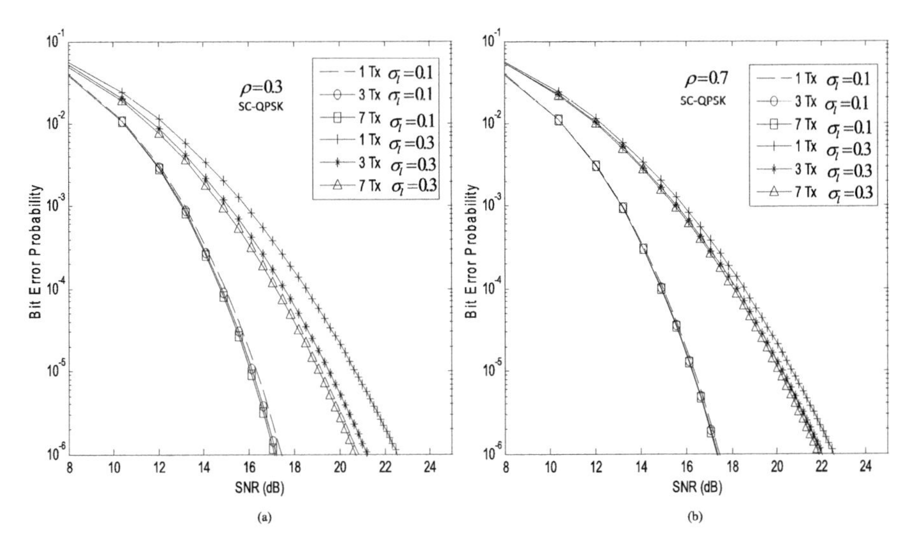

Fig. 6.14 Bit error probability vs. SNR with spatial diversity in weak turbulence (σ_l = 0.1 and 0.3) for SC-QPSK (**a**) $\rho = 0.3$ and (**b**) $\rho = 0.7$

(v) At the same diversity level, improvement in the performance decreases with the increase in the correlation coefficient for both SC-BPSK and SC-QPSK schemes.

6.4 Coding

Channel coding is another approach to achieve effective and reliable communication through the FSO turbulent channel. Channel capacity is one of the most important measure of communication channel which gives the highest information rate that can be reliably communicated through the channel. Channel coding aims to approach channel capacity promised by information theory and to achieve reliable information transmission. Lots of research has been carried out to determine the capacity of "classical optical channel," i.e., in the absence of atmospheric turbulence. The earliest works have considered a Poisson channel model for the quantum-noise limited receivers, assuming negligible thermal and background noise. It was observed that in photon counting receivers under average optical power limited condition, $\mathbb{M}$-PPM can help in achieving better performance in terms of BER. However, under an additional constraint of fixed peak optical power, binary level modulation schemes are giving better results for channel capacity. It is well known that PPM-based photon counting schemes cause increase in bandwidth with the increase in data rate. Therefore, in order to accommodate more data rate without increase in bandwidth, variants of PPM or pulse amplitude modulation (PAM) can provide better options.

FSO communication systems are mostly power limited due to limited power supply and safety regulations specially for long distance communication. Therefore, in order to overcome power loss in fading environment, besides selection of modulation scheme, a good channel coding technique is also used to mitigate the effect of turbulence in the atmosphere.

6.5 Channel Capacity

The capacity of a channel is the upper bound on the amount of information that can actually be transmitted reliably on the communication channel. Let X represents the set of signals that can be transmitted and Y the received signal, then the conditional distribution function of Y given X is $P_{Y/X}(y/x)$. The joint distribution of X and Y can be determined using the identity

$$P_{XY}(x, y) = P_{Y/X}(y/x) P_X(x) \tag{6.45}$$

where $P_X(x)$ is the marginal distribution of X and is given by

$$P_X(x) = \int_y P_{XY}(x, y)\, dy \tag{6.46}$$

Under these constraints, the maximum amount of information that can be communicated over the channel is the mutual information $I(X; Y)$, and this maximum mutual information is called the channel capacity, i.e.,

$$C = \max I(X; Y) \tag{6.47}$$

where $I(X; Y) = \sum_{y \in Y} \sum_{x \in X} P_{XY}(x, y) \log\left(\frac{P_{XY}(x,y)}{P_X(x) P_y(y)}\right)$. For continuous random variables, $I(X; Y)$ is given as $I(X; Y) = \int_Y \int_X P_{XY}(x, y) \log\left(\frac{P_{XY}(x,y)}{P_X(x) P_Y(y)}\right) dxdy$. For any given channel state $h(t)$, the average channel capacity can be determined as

$$\bar{C} = \int_{h=0}^{\infty} f(h) \cdot C dh \tag{6.48}$$

The channel capacity of band-limited channel in the presence of additive white Gaussian noise from Shannon Hartley theorem is given as

$$C = B \log_2\left(1 + \frac{S}{N}\right) \tag{6.49}$$

where C is the capacity of channel in bits per second, B the bandwidth of channel in Hertz, and S/N the signal to noise ratio. The capacity of the channel can also be expressed in terms of bandwidth efficiency (bits/sec/Hz) if the frequency response of the channel is known. When the information rate $R_b \leq C$, one can approach arbitrarily small bit error probabilities at a given SNR by using intelligent error-correcting coding techniques. For lower bit error probabilities, the encoder has to work on longer blocks of signal data. However, this entails longer delays and higher computational requirements. If information rate $R_b > C$, then errors cannot be avoided regardless of the coding technique used.

Shannon capacity may be thought of as the capacity of the channel with constraints on the input but no distortion from the channel. It gives the (theoretical) maximum data rate that can be transmitted with an arbitrarily small BER over a channel for a given average signal power. Therefore, in a channel capacity expression, if channel losses are reduced to zero, it will yield Shannon capacity. The channel capacity in case of FSO channel is considered as random variable due to randomness of the atmospheric channel. For such random channels subject to fading, the definitions of ergodic (average) or outage capacities are used and is defined as the expectation of the instantaneous channel capacity. Ergodic capacity is useful in case of fast fading channels, i.e., when the channel varies very fast with respect to the symbol duration. However, in case of slow fading channel, i.e., the channel coherence time is relatively large, the outage capacity becomes more meaningful. In this case, communication is declared successful if the mutual information exceeds the information rate. Otherwise, an outage event is declared which is commonly called probability of fade or outage probability.

6.5.1 Channel Coding in FSO System

Channel coding or forward error correction is a technique for controlling errors in data transmission over noisy channel by adding redundancy in the transmitted information. A block diagram of FSO communication system with encoder and decoder is shown in Fig. 6.15. The input data U are first encoded by an error-correcting code and are then mapped to coded bits C. In the encoding process, redundant information is added to the input data to aid in correcting the errors in the received signal. These coded bits C are then passed through the modulator which maps the coded bits into symbols X. The various set of symbols X represent the

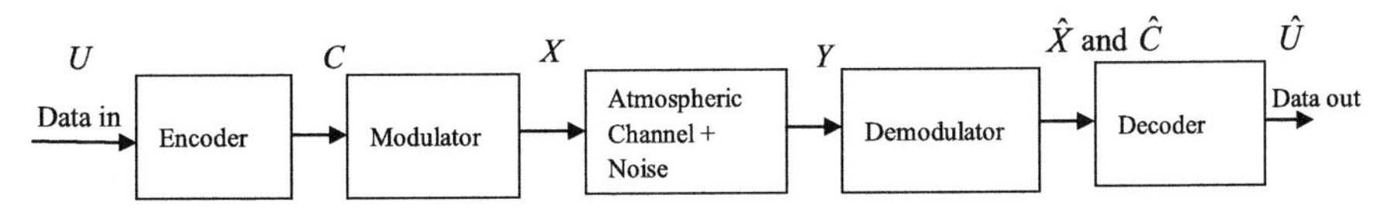

Fig. 6.15 Block diagram of FSO communication system with encoder and decoder

distinct messages that the laser will transmit. These symbols are then transmitted over the turbulent and noisy atmospheric channel, and the received set of symbols is referred by Y. At the receiver, demodulation and estimation are performed that accept the received symbols Y and produce the estimates of the transmitted symbols $\hat{X}$ or the coded symbols $\hat{C}$ or both. The decoder operates on these estimates to yield estimates of the data output $\widehat{U}$.

The error-correcting code is used to improve the data reliability by carefully adding some extra bits to the data bits that is to be transmitted through the channel. The process of adding extra bits is known as channel coding. Convolutional coding and block coding are the two major channel coding techniques that are discussed in this chapter. Convolutional codes operate on serial data, one or a few bits at a time. The encoders of convolutional codes are device with memory that accepts binary symbols in sets of k_b and outputs binary symbols in sets of n_c where every output symbol is determined by current input set as well as preceding input set. Block codes operate on relatively large message blocks (typically, up to a couple of hundred bytes). Also, encoder of the block code is a memoryless device which maps a k_b bit of sequence of input information in n_c coded output sequence. There are a variety of useful convolutional and block codes and various algorithms for decoding the received coded information sequences to recover the original data. The general terms used in coding are:

k_b: Number of "information" or "data" bits
m: Number of memory registers, and
n_c: Length of code ($n_c = m + k_b$)

There are two types of decoding: soft decision and hard decision decoding. In hard decision decoding, the received pulses are compared with a single threshold. If the voltage of the received signal is more than the threshold, it is considered to be one regardless of how close it is to the threshold; otherwise, it is definitely zero. In the case of soft decision decoding, the received signal is compared with various signal points in the constellation of the coded modulation system, and the one with minimum Euclidean distance is chosen. The optimum signal detection scheme on AWGN channel is based on minimizing the Euclidean distance.

6.5.1.1 Convolutional Codes

Convolutional codes have played an important role in communication systems including FSO links. These codes have two main parameters, i.e., code rate (k_b/n_c) and constraint length $\mathbb{L}$. The code rate k_b/n_c is a measure of efficiency of the code and is defined as the ratio of number of bits input to the encoder k_b to the number of symbols output of the encoder n_c. The constraint length $\mathbb{L}$ represents the number of bits in the encoder memory that affect the generation of the n_c output bits. In general, constraint length $\mathbb{L}$ is given by $\mathbb{L} = k_b(m-1)$. The encoded signal from convolutional encoder can be decoded either using Viterbi decoding or sequential decoding. The later decoding has the advantage that it can perform

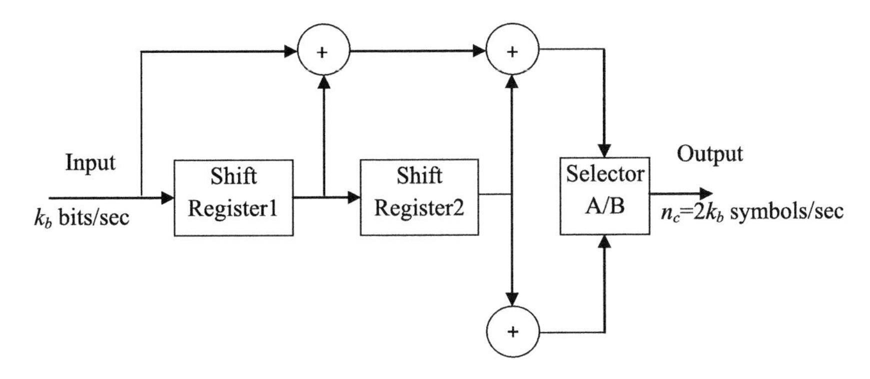

Fig. 6.16 Convolutional encoder with two memory elements and code rate = 1/2

very well with long constraint length, but it has a variable decoding time. Viterbi decoding has the advantage that it has a fixed decoding time. It is well suited to hardware implementation of the decoder, but its computational requirements grow exponentially as a function of the constraint length.

The convolutional code is generally represented as (n_c, k_b, m). The structure of convolutional encoder for rate 1/2 having constraint length 3 is shown in Fig. 6.16. The k_b bits/s is input to the encoder and it produces $2k_b$ symbols/s. The encoding of the data is accomplished using a shift register and associated combinatorial logic that performs modulo-2 addition. The output selector A/B cycles through two states: in the first state, it selects and gives the output of the upper modulo-2 adder, and in the second state, it selects and gives the outputs of the lower modulo-2 adder. This encoded signal after modulation is then transmitted through the noisy and turbulent atmospheric channel.

At the receiver side, the signal is demodulated and then the decoding is carried out using Viterbi decoder. The goal of the Viterbi algorithm is to find the transmitted sequence (or codeword) that is closest to the received sequence. An ideal Viterbi decoder would work with infinite precision. The received channel symbols are quantized with one or a few bits of precision in order to reduce the complexity of the Viterbi decoder. If the received channel symbols are quantized to one-bit precision (<0 V for "0" and ≥ 0 V for "1"), the result is called hard decision data. If the received channel symbols are quantized with more than one bit of precision, the result is called soft decision data. A Viterbi decoder with soft decision where the data is quantized to three or four bits of precision can perform about 2 dB better than one working with hard decision inputs [16]. The Viterbi algorithm also uses the trellis diagram to compute the accumulated distances (called the path metrics) from the received sequence to the possible transmitted sequences. The total number of such trellis paths grows exponentially with the number of stages in the trellis, causing potential complexity and memory problems like size, delays, etc.

The coding gain for FSO channel has been obtained for coherent (SC-BPSK and SC-QPSK) schemes in weak atmospheric turbulent environment ($\sigma_l = 0.25$) using Monte Carlo simulations. The simulations have been carried out with the following

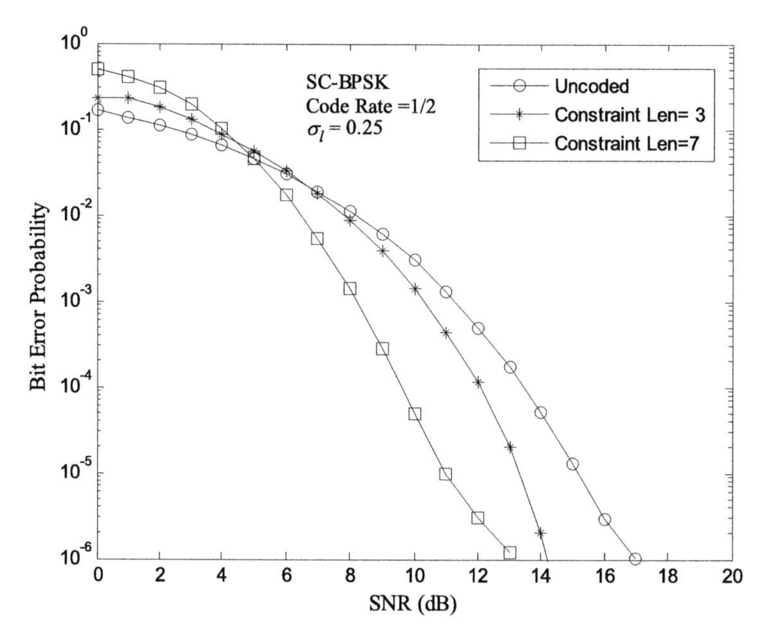

Fig. 6.17 Bit error probability for SC-BPSK with convolutional code ($\mathbb{L} = 3$ and 7) and code rate = 1/2

criteria for stopping the simulation program: maximum number of processed bits $= 10^6$ and maximum number of bit errors = 30. Subsequently, improvement in system performance in terms of coding gain is determined from the simulation results.

The BER variations with SNR obtained from Monte Carlo simulations for SC-BPSK and SC-QPSK using convolutional codes with constraint lengths 3 and 7 are shown in Figs. 6.17 and 6.18, respectively. It is observed that there is an improvement in the performance with the usage of convolutional code of constraint lengths 3 and 7 at high SNR. However, there is a degradation in the performance at lower SNR (0–5 dB). There is a crossover point in the performance of coded and uncoded systems. This crossover point occurs early with the increase in the constraint length. For example, the crossover point occurs at 7 dB when the constraint length is 3 for SC-BPSK. This crossover point shifts to 5 dB for constraint length 7. The physical reason for this crossover is that certain minimum number of bits are required by the decoder at the receiver to achieve the advantage due to coding. Until minimum number of bits are there, performance of coded system will be worse than the uncoded system. Convolutional coding with larger constraint length enables more number of bits to be processed at a given time, and hence its crossover point comes early than that with small constraint length.

When a comparison is made between uncoded and coded (convolutional code with constraint length $\mathbb{L} = 3$ and 7) systems, it is observed that at lower SNR, the performance with $\mathbb{L} = 7$ is poorer than with $\mathbb{L} = 3$. At the same time, there will be increase in processing time for $\mathbb{L} = 7$. However, at higher SNR, the performance

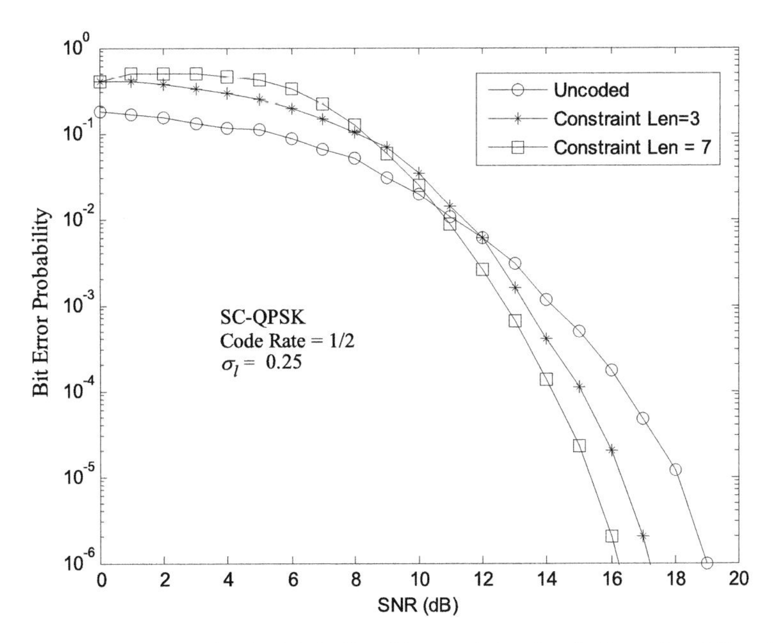

Fig. 6.18 Bit error probability for SC-QPSK with convolutional code ($\mathbb{L} = 3$ and 7) and code rate = 1/2

is better with $\mathbb{L} = 7$ than that with $\mathbb{L} = 3$. At the higher constraint length, the disadvantage of increase in processing time is still there. It is therefore inferred that at lower SNR, convolutional code with smaller constraint length and at higher SNR convolutional code with large constraint length are preferred.

6.5.1.2 Low Density Parity Check Codes

Low density parity check (LDPC) codes are a class of linear block LDPC codes. These codes are constructed with the help of the parity check matrix, which is sparse, i.e., it contains only a few 1s in comparison to the number of 0s. LDPC codes if properly designed has the ability to give the performance up to Shannon limit. The performance of LDPC code depends upon the complexity of the decoding algorithm which is directly related to the density of 1s in the matrix. A parity check matrix for a LDPC code with dimension $m \times n$ is given by

$$H = \begin{bmatrix} 1\,1\,1\,1\,0\,0\,0\,0\,0\,0 \\ 1\,0\,0\,0\,1\,1\,1\,0\,0\,0 \\ 0\,1\,0\,0\,1\,0\,0\,1\,1\,0 \\ 0\,0\,1\,0\,0\,1\,0\,1\,0\,1 \\ 0\,0\,0\,1\,0\,0\,1\,0\,1\,1 \end{bmatrix}_{m \times n} \tag{6.50}$$

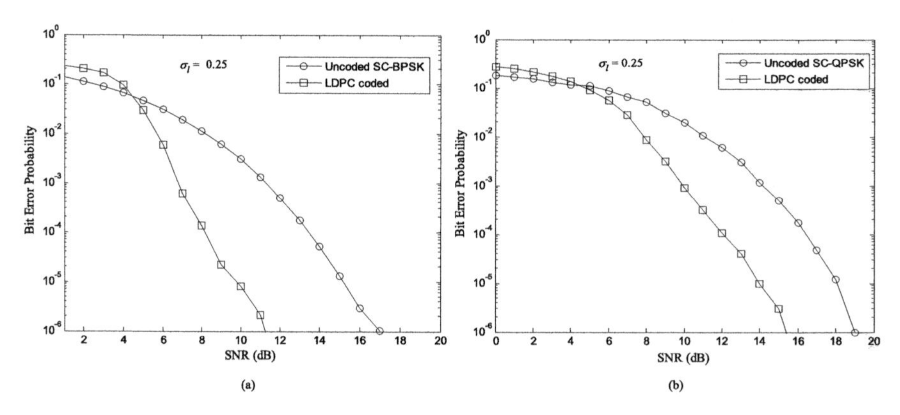

Fig. 6.19 Bit error probability with LDPC code at $\sigma_l = 0.25$ for (**a**) SC-BPSK and (**b**) SC-QPSK

Let w_r represents the number of 1s in each row and w_c number of 1s in each column. For a matrix to be called low density, the two conditions $w_c < m$ and $w_r < n$ must be satisfied where m and n are number of rows and columns, respectively. A regular LDPC code has $w_c n = w_r m =$ number of 1s in H matrix and code rate $= k_b/n_c \geq 1-(w_c/w_r)$. Such codes are therefore referred as (w_c, w_r) regular LDPC codes. If H is low density but the numbers of 1s in each row or column are not constant, the code is called a irregular LDPC code. Therefore, Eq. (6.50) is (2, 4) regular LDPC code with $n_c = 10, m = 5, w_c = 2$ and $w_r = 4$.

Several different algorithms exist to construct suitable LDPC codes. One way is to semi-randomly generate sparse parity check matrices [17]. Codes constructed by this approach are good; however, the encoding complexity of such codes is usually high. The decoding of LDPC codes is done using iterative decoding algorithms. These algorithms perform local calculations and this step is typically repeated several times. The term local calculations implies that a divide and conquer strategy is used, which separates a complex problem into manageable subproblems. A sparse parity check matrix now helps conquer this algorithms in several ways. It helps to keep both the local calculations simple and also reduces the complexity of combining the subproblems by reducing the number of needed messages to exchange all the information. Further, it is reported that iterative decoding algorithms of sparse codes perform very close to the optimal maximum likelihood decoder [18].

The Monte Carlo simulations results for BER performance with LDPC code for SC-BPSK and SC-QPSK are shown in Fig. 6.19a, b, respectively.

The coding gains of convolutional code with constraint lengths 3 and 7 over uncoded system for a given BER (10^{-6} and 10^{-4}) are obtained, and the results are presented in Table 6.4. In the same table, the results for LDPC code from Fig. 6.19a, b are also given.

Following observations are made from Figs. 6.17, 6.18, and 6.19 and Table 6.4 for coherent modulation schemes, viz., SC-BPSK and SC-QPSK:

Table 6.4 Comparison of coding gains with convolutional and LDPC codes for SC-BPSK and SC-QPSK modulation schemes in weak atmospheric turbulence ($\sigma_l = 0.25$) at BER = 10^{-6} and 10^{-4}, respectively

Modulation scheme	Coding gain (in dB) at BER = 10^{-6}			Coding gain (in dB) at BER = 10^{-4}		
	Convolutional code			Convolutional code		
	$\mathbb{L} = 3$	$\mathbb{L} = 7$	LDPC	$\mathbb{L} = 3$	$\mathbb{L} = 7$	LDPC
SC-BPSK	2.8	4.0	5.7	1.5	3.8	5.3
SC-QPSK	1.7	2.7	3.5	1.3	2.2	3.3

(i) It is seen that the SC-BPSK gives better performance than SC-QPSK for both convolutional and LDPC coding schemes.
(ii) Improvement in coding gain (in dB) is more with the increase in constraint length for convolutional codes at higher SNR.
(iii) The coding gain is more for LDPC code as compared to convolutional code for both SC-BPSK and SC-QPSK modulation schemes.
(iv) The above observations are valid at both BER = 10^{-6} and 10^{-4}. However, the coding gain is relatively more at lower BER. This is true for both the modulation schemes.

6.6 Adaptive Optics

Adaptive optics (AO) is a technique that is used to correct the optical wave front by providing atmospheric compensation that is based on wave front sensing and reconstruction. This technique pre-corrects the transmitted beam before propagating it into the atmosphere and therefore reduces the signal fades caused by scintillation. There are two important components used in adaptive optics system: (i) "deformable mirror" which actually makes the optical corrections and the "wave front sensor" which measures the turbulence hundreds of times a second. These are connected together by a high-speed computer.

The deformable mirrors are made up of thin sheet of glass, and this glass is attached to the actuators (devices which expand or contract in length in response to a voltage signal, bending the thin sheet of glass locally) at their back. These days, deformable mirrors are based on microelectromechanical system (MEMS) using small piezoelectric actuators. Wave front sensors make use of detectors that are either charge-coupled devices (CCD) or set of avalanche photodiodes.

Conventional adaptive optics systems are based on the wave front conjugation principle. Phase conjugation is realized in an optoelectronic feedback loop system comprising of wave front sensor and reconstructor, control system, and deformable mirror as shown in Fig. 6.20. In order to avoid wave front measurements that are not desired in strong turbulent conditions, control of wave front correctors in AO system can be performed by blind or model-free optimization. A part of the received

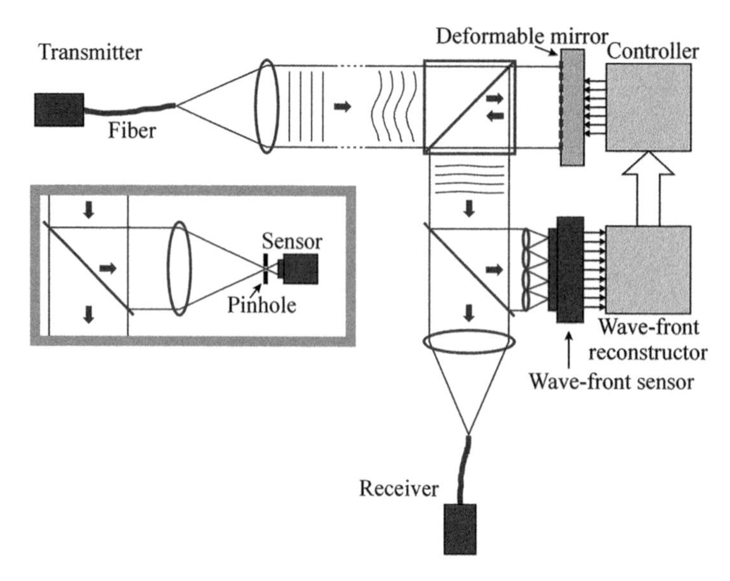

Fig. 6.20 Conventional adaptive optics system using wave front sensor and reconstructor

signal is sent to wave front sensor that produces a control signal for the deformable mirror as shown in Fig. 6.21. However, this method of wave front control imposes a serious limitation for the control bandwidth. Model-free optimization in real-time adaptive optics systems imposes the difficulty to control a deformable mirror with a number of controllable elements (up to the order of magnitude of hundreds) with a single scalar feedback signal. The information necessary to control individual mirror elements has to be acquired in time or frequency domain from a single sensor rather than from parallel signals of spatially distributed sensor arrays. Thus, the bandwidth of deformable mirror and controller required for real-time correction of atmospheric wave front distortion was beyond the limitation of available hardware. However, recent development in micromachined deformable mirrors (μDMs) has shown model-free AO system a prospective.

Adaptive optics based on model-free optimization can be used in different ways in FSO communication system: AO receiver, AO transmitter, or AO transceiver architectures. AO receiver is most commonly used where the deformations in the received beam are compensated by AO system allowing a better focusing of the optical signal onto the small receiver aperture area. However, the potential of AO receiver in FSO is very limited as it compensates only for the light waves that have entered the receiver aperture. In order to improve the performance of system, AO transmitter system is essential at the other end of the propagation path as it allows for per-compensation of the transmitted beam and helps in mitigating turbulence-induced beam spreading.

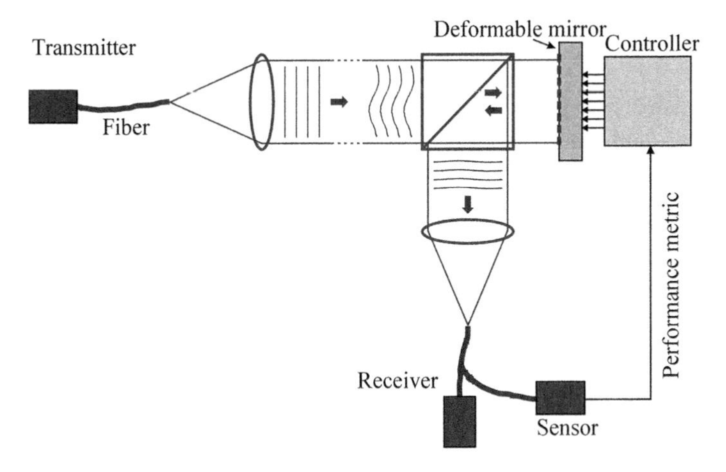

Fig. 6.21 Model free adaptive optics system

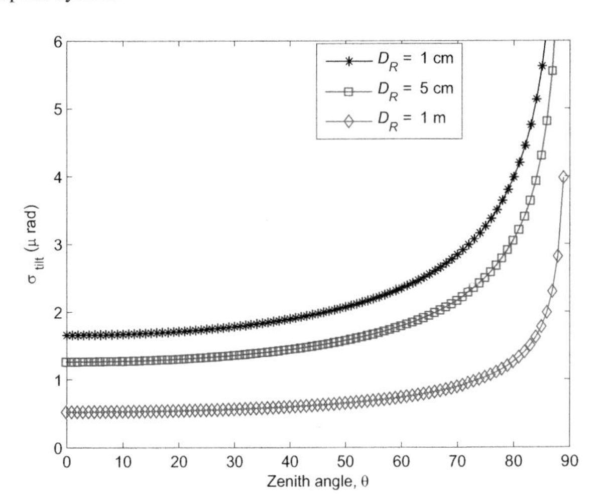

Fig. 6.22 RMS wave front tilt as a function of zenith angle for different telescope apertures

The rms atmospheric turbulence-induced wave front tip/tilt is given as

$$\sigma_{tilt} = 0.43\left(\frac{\lambda}{D_R}\right)\left(\frac{D_R}{r_0}\right)^{5/6} \tag{6.51}$$

It is clear from Eq. (6.51) that wave front tilt is independent of wavelength as coherence length r_0 is a function of wavelength. Figure 6.22 shows the rms wave front tilt as a function of zenith angle for different telescope apertures. It is clear from the figure that rms tilt error increases with zenith angle. For a given zenith angle, rms tilt is smaller for larger telescope aperture.

6.7 Relay-Assisted FSO Transmission

Relay-assisted transmission is an alternative way of realizing spatial diversity advantages, and it is a powerful mitigation tool in the presence of atmospheric turbulence [19, 20]. This technique was earlier used with RF technology and is now studied in context with FSO communication wherein the diversity gain can compensate for the loss in fading channels leading to improved FSO links. The main idea behind cooperative diversity is based on the fact that the broadcasted wireless RF signals are overheard by other nodes, which are also called partners or relays. So, cooperative diversity makes use of this overheard information for transmission from source to destination via relays. Hence, a simple cooperative network will consist of three components/nodes, i.e., source, destination, and relay where each node is having only one antenna. If the distance between the source and destination is too large (in the order of kms), then the overheard information by the relay node is transmitted to the destination via different paths following either series or parallel relay configurations as shown in Fig. 6.23.

These multiple single antenna nodes form a virtual antenna array realizing spatial diversity in a distributed fashion. These relays can be further classified in four ways: (i) decode-and-forward (DF), (ii) compress-and-forward (CF), (iii) amplify-and-forward (AF), and (iv) detect-and- forward (DEF). In DF relay, the relay decodes the source message and transmits the encoded message to the destination. In CF, the relay quantizes the received signal from the source and transmits the encoded version of the quantized received signal to the destination. In AF relay, the relay transmits the amplified version of the received signal from source to destination. AF relay are low-complexity relay transceivers since there is no signal processing involved for encoding and decoding process. However, main disadvantage of AF relays is that it will also forward noise which is received at the relay. In DEF, the relay detects the signal (hard decision detection), modulates it, and forwards it to the destination. Recently, adaptive DEF or adaptive DF has been proposed [21] where

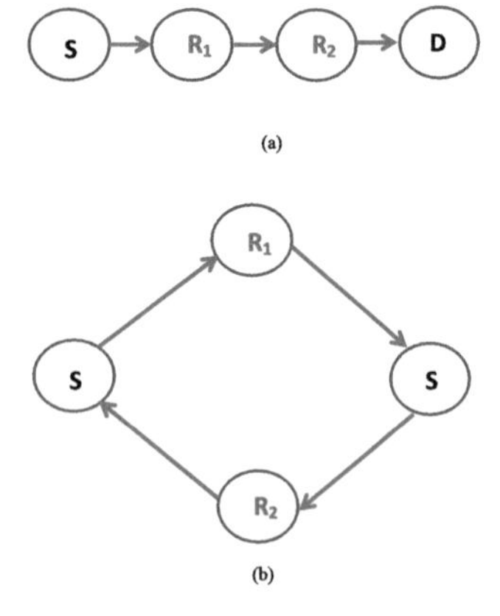

Fig. 6.23 Relay configurations: (**a**) series relay and (**b**) parallel relay

the relay takes part in the data transmission only if it can receive error-free data frames from the source or when the SNR at the relay is large enough, respectively.

Relay-assisted FSO transmission was first proposed by Acampora and Krishnamurthy in [22], where the performance of a mesh FSO network was investigated from a network capacity point of view. The work presented in this paper coves networking perspective and does not take into account communication through physical layer. Later, Karagiannidis et al. [23] have evaluated outage probability for a multi-hop FSO system using K and gamma-gamma atmospheric turbulent channel models without considering path loss. Multi-hop transmission is an alternative relay-assisted transmission scheme which employs the relays in a serial configuration (refer to Fig. 6.23a). It is observed that multi-hop serial relay FSO transmission provides range extension and good diversity gain. The diversity gain in lognormal turbulence channels (assuming plane wave propagation) gives the diversity order of $(Z+1)^{11/6}$ where Z is the number of relays. Relay-assisted parallel FSO transmission (i.e., cooperative diversity) shown in Fig. 6.23b is considered where line-of-sight FSO through directional beams is not possible. It makes use of multiple transmitting laser pointing in the direction of corresponding relay nodes. In this case, the source node transmits the same information to Z relays, and these relays further retransmit the information to the destination. For parallel relaying, all relays should be located at the same place (along the direct link between the source and the destination) closer to the source, and the exact location of this place turns out to be a function of SNR, the number of relays, and the end-to-end link distance [24]. It is to be noted that relay-assisted FSO transmission does not require distributed space-time block codes like in RF transmission as the received signals ensures orthogonality from sufficiently separated transmit apertures [25].

An improvement in relay-assisted FSO link for a certain scenario has been studied in [26]: weak atmospheric turbulence with $C_n^2 = 10^{-14}\ \mathrm{m}^{-2/3}$, link length =5 km, atmospheric attenuation =0.43 dB/km, and target outage probability $=10^{-6}$. It is seen that improvements of 18.5 and 25.4 dB in power margin are achieved when one or two equidistant relays are placed between source and destination using series DF relay-assisted FSO transmission. When AF relay transmission is used, the improvements are about 12.2 and 17.7 dB, respectively. In case of parallel relay FSO transmission (relay are placed midway between source and destination), the obtained improvements are about 20.3 and 20.7 dB for DF mode, and 18.1 and 20.2 dB for AF mode, for the cases of two and three relays, respectively.

6.8 Summary

This chapter discusses various performance improvement techniques that help to mitigate the adverse effect of the atmosphere. It starts with aperture averaging wherein it is observed that the effect of turbulence-induced scintillation is decreased with the increase in the receiver aperture size. However, the receiver aperture size cannot be increased indefinitely as it increases the cost as well as background noise. So other alternative approach is diversity technique which is an effective approach to improve the performance of an FSO system by providing redundancy across spatially separated branches. It is observed that maximum diversity gain is achieved when the correlation between the branches is zero, i.e., branches are independent. Alamouti transmit diversity is also discussed that utilizes both time and space diversity, and it provides diversity order similar to that of MRC receiver. Error-correcting codes provide a reliable communication approach in noisy environment. It increases the channel bandwidth requirement by adding redundant bits in the coded data, but it makes the system power efficient with high data rates. The performance of two error-correcting codes, i.e., convolutional and LDPC codes, is discussed. It is observed that LDPC codes provide large SNR gain for same code rate. LDPC codes offer low encoder and decoder design complications and hence are good choice for a practical FSO communication system. Adaptive optics system has been discussed briefly that makes use of wave front sensors and deformable mirrors to reduce the signal fading due to atmospheric turbulence. Relay-assisted FSO transmission using multi-hop series transmission and cooperative diversity has been presented as a fading mitigation tool in the presence of atmospheric turbulence.

Bibliography

1. J. Churnside, Aperture averaging of optical scintillation in the turbulent atmosphere. Appl. Opt. **30**(15), 1982–1994 (1991)
2. L.C. Andrews, R.L. Phillips, *Laser Beam Progapation Through Random Media*, 2nd edn. (SPIE Press, Bellingham, Washington, 2005)
3. M. Khalighi, N. Schwartz, N. Aitamer, S. Bourennane, Fading reduction by aperture averaging and spatial diversity in optical wireless systems. J. Opt. Commun. Net. **1**(6), 580–593 (2009)
4. L.C. Andrews, R.L. Phillips, C.Y. Hopen, Aperture averaging of optical scintillations: power fluctuations and the temporal spectrum. Waves Random Media **10**(1), 53–70 (2000)
5. J. Churnside, Aperture averaging factor for optical propagation through turbulent atmosphere, in *NOAA Technical Memorandum ERLWPL-188* (1990). [Weblink: http://www.ntis.gov]
6. L.C. Andrews, Aperture averaging factor for optical scintillation of plane and spherical waves in the atmosphere. J. Opt. Soc. Am. **9**(4), 597–600 (1992)
7. H.T. Yura, W.G. McKinley, Aperture averaging for space-to-ground optical communications applications. Appl. Opt. **22**(11), 1608–1609 (1983)
8. R.J. Hill, S.F. Clifford, Modified spectrum of atmospheric temperature fluctuations and its applications to optical propagation. J. Opt. Soc. Am. **68**(7), 892–899 (1978)
9. M. Khalighi, N. Aitamer, N. Schwartz, S. Bourennane, Turbulence mitigation by aperture averaging in wireless optical systems, in *Proceedings IEEE, International Conference on Telecommunication – ConTel 2009*, Zagreb (2009), pp. 59–66
10. H. Kaushal, V. Kumar, A. Dutta, A. Aennam, V.K. Jain, S. Kar, J. Joseph, Experimental study on beam wander under varying atmospheric turbulence conditions. IEEE Photon. Technol. Lett. **23**(22), 1691–1693 (2011)
11. N. Mehta, H. Kaushal, V.K. Jain, S. Kar, Experimental study on aperture averaging in free space optical communication link, in *National Conference on Communication-NCC*, 15th–17th Feb 2013

12. S.M. Alamouti, A simple transmit diveristy technique for wireless communications. IEEE J. Select. Areas Commun. **16**(8), 1451–1458 (1998)
13. Q. Shi, Y. Karasawa, An accurate and efficient approximation to the Gaussian Q-function and its applications in performance analysis in Nakagami-m fading. IEEE Commun. Lett. **15**(5), 479–481 (2011)
14. Q. Liu, D.A. Pierce, A note on Gauss-Hermite quadrature. J. Biom. **81**(3), 624–629 (1994)
15. X. Zhu, J.M. Kahn, Free space optical communication through atmospheric turbulence channels. IEEE Trans. Commun. **50**(8), 1293–1300 (2002)
16. O.O. Khalifa, T. Al-maznaee, M. Munjid, A.A. Hashim, Convolution coder software implementation using VIiterbi decoding algorithm. J. Comput. Sci. **4**(10), 847–856 (2008)
17. D.J.C. MacKay, Good error-correcting codes based on very sparse matrices. IEEE Trans. Inf. Theory **45**(2), 399–431 (1999)
18. L. Yang, M. Tomlinson, M. Ambroze, J. Cai, Extended optimum decoding for LDPC codes based on exhaustive tree search algorithm, in *IEEE International Conference on Communication Systems*-ICCS (2010), pp. 208–212
19. A. Nosratinia, T.E. Hunter, A. Hedayat, Cooperative communication in wireless networks. IEEE Commun. Mag. **42**(10), 74–80 (2004)
20. J.N. Laneman, D.N.C. Tse, G.W. Wornell, Cooperative diversity in wireless networks: efficient protocols and outage behavior. IEEE Trans. Inf. Theory **50**(12), 3062–3080 (2004)
21. M. Karimi, N. Nasiri-Kenari, BER analysis of cooperative systems in free-space optical networks. J. Lightw. Technol. **27**(12), 5639–5647 (2009)
22. A. Acampora, S. Krishnamurthy, A broadband wireless access network based on mesh-connected free-space optical links. IEEE Pers. Commun. **6**(10), 62–65 (1999)
23. G. Karagiannidis, T. Tsiftsis, H. Sandalidis, Outage probability of relayed free space optical communication systems. Electron. Lett. **42**(17), 994–996 (2006)
24. M.A. Kashani, M. Safari, M. Uysal, Optimal relay placement and diversity analysis of relay-assisted free-space optical communication systems. IEEE J. Opt. Commun. Netw. **5**(1), 37–47 (2013)
25. M. Safari, M. Uysal, Do we really need OSTBC for free-space optical communication with direct detection? IEEE Trans. Wirel. Commun. **7**, 4445–4448 (2008)
26. M. Safari, M. Uysal, Relay-assisted free-space optical communication. IEEE Trans. Wirel. Commun. **7**(12), 5441–5449 (2008)

Chapter 7
Link Feasibility Study

7.1 Link Requirements and Basic Parameters

For the link design, the initial link requirements, viz., data rate, bit error rate, acquisition time, and range, are to be specified. For our study in this chapter, we have considered typical requirements of bit error rate less than one error in a million bits transferred, i.e., $P_e \leq 10^{-6}$. The link design requirements are given in Table 7.1.

Keeping in view the complexity and other constraints, three-level transmit diversity using SC-BPSK modulation is considered for subsequent calculations of link margin. For this diversity level, the improvement in link margin has been worked out using LDPC coding scheme. In making the link power budget, commonly used parameters along with their abbreviations are given in Table. 7.2.

The received optical signal power at the photodetector input in the receiver is calculated from the range equation which is given as

$$P_R = P_T G_T \eta_T \eta_{TP} \left(L_s\right) G_R \eta_R \eta_\lambda \tag{7.1}$$

where P_T and P_R are the transmitter and receiver optical power, respectively; G_T and G_R the transmitter and receiver antennae gain, respectively; η_T and η_R the transmitter and receiver optic efficiency, respectively; L_s the space loss factor; η_{TP} the pointing loss; and η_λ the narrow band filter (NBF) transmission loss. In the above range equation, the first four parameters (P_T, G_T, η_T, and η_{TP}) are the transmitter parameters, the fifth parameter is the atmospheric transmission loss (L_s), and the last three parameters (P_R, G_R, η_R) are the receiver parameters. These parameters are discussed in the following subsections.

H. Kaushal et al., *Free Space Optical Communication*, Optical Networks,
DOI 10.1007/978-81-322-3691-7_7

Table 7.1 Link design requirements

Requirement	Value
Data rate	500 Mbps
Bit error rate	$\leq 10^{-6}$
Range	40,000 km

Table 7.2 Commonly used parameters and their abbreviations in link power budget

Parameters	Abbreviations
Laser power	P_T
Laser operating wavelength	λ
Transmit telescope aperture diameter	D_T
Transmit telescope obscuration ratio	γ_T
Transmitter optics throughput	η_T
Transmitter pointing loss factor	η_{TP}
Receiver telescope aperture diameter	D_R
Receiver telescope obscuration ratio	γ_R
Receiver optics throughput	η_R
Narrow band filter spectral bandwidth	$\Delta\lambda_{filter}$
Narrow band filter transmission loss	η_λ

7.1.1 Transmitter Parameters

The parameters $G_T\eta_T$ together give the net on-axis gain of the transmitter. The parameter G_T is the gain of the telescope and is given by

$$G_T = \left(\frac{16}{\theta_{div}^2}\right) \tag{7.2}$$

The angular divergence θ_{div} is directly proportional to wavelength λ and inversely proportional to the transmitter aperture diameter D_T, i.e.,

$$\theta_{div} = \frac{4\lambda}{\pi D_T} \tag{7.3}$$

Substitution of Eq. (7.3) in Eq. (7.2) leads to

$$G_T = \left(\frac{\pi D_T}{\lambda}\right)^2 = \left(\frac{4\pi A}{\lambda^2}\right) \tag{7.4}$$

where A $(= \pi D_T^2/4)$ is the aperture area. The on-axis gain of a Gaussian beam with a central obscuration [1] is given as

$$G_T = \left(\frac{4\pi A}{\lambda^2}\right)\left[\frac{2}{\alpha_T^2}\left\{e^{-\alpha_T^2} - e^{-\alpha_T^2\gamma_T^2}\right\}^2\right] \tag{7.5}$$

where α_T is the truncation ratio (defined as the ratio of the main aperture diameter to the Gaussian beam spot size, i.e., D_T/D_{spot}). It is given by

$$\alpha_T = 1.12 - 1.30\gamma_T^2 + 2.12\gamma_T^4 \tag{7.6}$$

In the above equation, the parameter γ_T is the obscuration ratio (defined as the ratio of the central obscuration diameter to the main aperture diameter). It may be mentioned that Eq. (7.6) is valid for $\gamma_T \leq 0.4$. In the limiting case, when there is no obscuration, i.e., $\gamma_T = 0$, Eq. (7.5) reduces to

$$G_T = \left(\frac{4\pi A}{\lambda^2}\right)\left[\frac{2}{\alpha_T^2}\left\{e^{-\alpha_T^2} - 1\right\}^2\right] \tag{7.7}$$

The off-axis gain of the Gaussian beam is approximately given by

$$G_T\,(off - axis) \simeq \left(\frac{4\pi A}{\lambda^2}\right) e^{-8\left(\theta_{off}/\theta_{div}\right)^2} \tag{7.8}$$

where θ_{off} is the off-axis angle and θ_{div} the $1/e^2$ of beam diameter. In our link power budget calculations, we have considered only the on-axis gain G_T.

The parameter η_T in Eq. (7.1) is the transmitter optic efficiency that takes into account the transmission and reflection losses in the transmitter, i.e., those in the relay optics, in the steering mirrors, and in the telescope. Its typical value is in the range of 0.4 to 0.7 depending upon transmission and reflection coefficients of the optical components in the transmitting system. In the link power budget calculations, we have taken the value of η_T to be 0.65. The term η_{TP} in Eq. (7.1) is the transmitter pointing loss factor. In most cases, the mean value of this factor [2, 3] is considered and is given by

$$\eta_{TP} = \int_0^\infty \eta_{TP}\,(\alpha_r)\,\frac{\alpha_r}{\sigma_T^2}\exp - \left(\frac{\alpha_r^2 + \epsilon_T^2}{2\sigma_T^2}\right) I_0\left(\frac{r\epsilon_T}{\sigma_T^2}\right) d\alpha_r \tag{7.9}$$

where α_r is the angular off-axis pointing displacement, ϵ_T the root sum square (RSS) of two axes pointing bias error, σ_T the RSS two axes jitter, I_0 the modified Bessel function of order zero, and $\eta_{TP}\,(\alpha_r)$ the instantaneous pointing loss as a function of angular off-axis pointing displacement. For a nominal pointing error, i.e., for $\alpha_r \leq \lambda/D_T$, $\eta_{TP}\,(\alpha_r)$ can be approximated by the series given as [4]

$$\eta_{TP}\,(\alpha_r) \cong \frac{1}{f_0^2\,(\gamma_T)}\left[f_0\,(\gamma_T) + \frac{f_2\,(\gamma_T)}{2!}x^2 + \frac{f_4\,(\gamma_T)}{4!}x^4 + \frac{f_6\,(\gamma_T)}{6!}x^6\right]^2 \tag{7.10}$$

where $x = \pi\,(D_T/\lambda)\,\alpha_r$ and the coefficients f_0, f_2, f_4, and f_6 are given in Table 7.3 for several values of γ_T. The angular off-axis pointing displacement is basically the pointing error. Since we have considered alignment error to be zero, therefore the

Table 7.3 Values of series coefficients for pointing loss factor calculation

Transmitter obscuration ratio, γ_T	f_0	f_2	f_4	f_6
0.0	0.569797	−0.113421	0.0503535	−0.0292921
0.1	0.566373	−0.115327	0.0513655	−0.0299359
0.2	0.555645	−0.120457	0.0542465	−0.0317773
0.3	0.535571	−0.126992	0.0584271	−0.0344978
0.4	0.501381	−0.131777	0.0626752	−0.0374276

pointing error will become the total pointing error. For angular off-axis pointing displacement $\alpha_r = 1\,\mu rad$, the condition $\alpha_r \le \lambda/D_T$ is satisfied for $D_T = 5.94$ cm and 9.42 cm at $\lambda = 1064$ nm and $\lambda = 1550$ nm, respectively. Therefore, in both the cases the pointing loss factor can be determined using Eq. (7.10) and Table 7.3. For a typical value of obscuration ratio $\gamma_T = 0.2$, the η_{TP} comes out to be nearly 0.9 at both 1064 and 1550 nm wavelengths. We have incorporated this loss factor in making the link power budget in Sect. 7.2.

7.1.2 Atmospheric Transmission Loss Parameter

Atmospheric transmission loss parameter L_s is basically the range loss (also called space loss). It is the largest loss in an FSO link when an optical beam is transversed through the link length/range R and is given by

$$L_s = \left(\frac{\lambda}{4\pi R}\right)^2 \tag{7.11}$$

Since the optical operating wavelengths are smaller than that used in a RF communication, the space loss incurred by an optical system is relatively much higher than in RF system.

7.1.3 Receiver Parameters

The term G_R is the gain of the receiver telescope, and it is calculated from the collecting area of the antenna and the operating wavelength. For an ideal receiving aperture with area equal to unobscured part of the telescope, the receiver gain is given as [4]

$$G_R = \left(\frac{\pi D_R}{\lambda}\right)^2 \left(1 - \gamma_R^2\right) \tag{7.12}$$

In the above equation, γ_R is the receiver obscuration ratio defined as the ratio of receiver obscuration diameter to the aperture diameter. The other receiver parameter η_R in the range equation (refer Eq. (7.1)) is the receiver optic efficiency that will take into account the transmission and reflection losses in the receiver. Its typical value ranges from 0.5 to 1. We have taken η_R to be 0.7 in the link power budget calculations. The last receiver parameter in range equation η_λ is NBF transmission loss. The typical value of η_λ is taken to be 0.7.

The narrow band filter is an important component in the optical communication system as it greatly affects the sensitivity and background noise rejection. Ideally, the filter should have 100 % transmission in the pass band and a very narrow spectral bandwidth $\Delta\lambda_{filter}$ (e.g., 1 Å or less). In the calculation of background noise power, we have taken $\Delta\lambda_{filter}$ to be 10 Å.

7.2 Link Power Budget

The link budget analysis for SC-BPSK modulation scheme with a given requirements (as in Table 7.1) and components/parameters values (as in Table 7.4) is given in Table 7.5.

It is observed from Table 7.5 that the link power margins at $\lambda = 1064$ and 1550 nm are 2.03 and 4.25 dB, respectively. These power margins are not adequate in the presence of absorption and scattering losses. In that case, we may like to use transmit diversity or LDPC codes. With transmit diversity, these power margins become 4.43 and 6.65 dB at $\lambda = 1064$ and 1550 nm, respectively. It implies an improvement of nearly 2.40 dB at both the wavelengths. Instead of diversity, if

Table 7.4 Various communication link components/parameters and their values for link power budget calculations

S.No.	Components/parameters	Value
1.	Laser power, P_T	3000 mW
2.	Operating wavelength, λ	1064 and 1550 nm
3.	Tx. Telescope diameter, D_T	5.94 and 9.42 cm
4.	Tx. Obscuration ratio, γ_T	0.2
5.	Tx. Optics efficiency, η_T	0.65
6.	Tx. Pointing loss factor, η_{TP}	0.9
7.	Rx. Telescope diameter, D_R	30 cm
8.	Rx. Obscuration ratio, γ_R	0.35
9.	Rx. Optics efficiency, η_R	0.7
10.	Spectral filter bandwidth, $\Delta\lambda_{filter}$	10 Å
11.	Data rate, R_b	500 Mbps
12.	Link range, R	40,000 km
13.	Zenith angle, θ	0°

Table 7.5 Link power budget of SC-BPSK modulation scheme using LDPC code for ground-to-satellite uplink at zero zenith angle

Transmitter Parameters					
S.No	**Parameters**	**Absolute Value**		**Equivalent Value (in dB or dBm)**	
		λ = 1064 nm	λ = 1550 nm	λ = 1064 nm	λ = 1550 nm
1.	Laser Power, P_T	3000 mW	3000 mW	34.77 dBm	34.77 dBm
2.	Tx. Telescope, G_T Tx. Obscuration Ratio, $\gamma_T = 0.2$ Tx. Telescope Diameter, $D_T = 5.92$ cm and 9.42 cm at λ = 1064 nm and 1550 nm, respectively.	3.13 x 10^{10}	3.64 x 10^{10}	105 dB	105.61 dB
3.	Tx. Optics, η_T	0.65	0.65	-1.87 dB	-1.87 dB
4.	Tx. Pointing Loss Factor, η_{TP}	0.9	0.9	-0.45 dB	-0.45 dB
Atmosphere Transmission Loss Parameter					
5.	Space Loss, L_s	4.48 x 10^{-30}	9.50 x 10^{-30}	-293.48 dB	-290.21 dB
Receiver Parameters					
6.	Rx. Telescope Gain, G_R Rx. Obscuration Ratio, $\gamma_T = 0.35$ Rx. Telescope Diameter, $D_R = 30$ cm	6.88 x 10^{11}	3.24 x 10^{11}	118.37 dB	115.11 dB
7.	Rx. Optics Efficiency, η_R	0.7	0.7	-1.54 dB	-1.54 dB
8.	NBF Transmission Loss, η_λ	0.7	0.7	-1.54 dB	-1.54 dB
9.	Received Signal Power, P_R (Using Rows 1 to 8)	8.97 x 10^{-8} W	9.72 x 10^{-8} W	-40.74 dBm	-40.12 dBm
10.	Minimum Required Rx. Signal Power using Single Transmitter, P_R(min)	5.24 x 10^{-8} W	3.63 x 10^{-8} W	-42.80 dBm	-44.39 dBm
11.	Minimum Required Rx. Signal Power using Three Transmitter, P_R(min)	2.96 x 10^{-8} W	2.05 x 10^{-8} W	-45.27 dBm	-46.86 dBm
12.	Minimum Required Rx. Signal Power with LDPC Codes, P_R(min)	9.97 x 10^{-9} W	6.91 x 10^{-9} W	-50.01 dBm	-51.59 dBm
13.	Link Margin using Single Tx. (Using Rows 9 and 10), P_R - P_R(min)	**1.59**	**2.66**	**2.03 dB**	**4.25 dB**
14.	Link Margin using Three Tx. (Using Rows 9 and 11), P_R - P_R(min)	**2.77**	**4.62**	**4.43 dB**	**6.65 dB**
15.	Link Margin with LDPC Codes (Using Rows 9 and 12), P_R - P_R(min)	**8.45**	**14.02**	**9.27 dB**	**11.47 dB**

LDPC code is used, then the corresponding improvement in the link power margins are 7.24 and 7.22 dB at $\lambda = 1064$ and 1550 nm, respectively. It is clear from above that usage of LDPC code instead of transmit diversity provides an additional gain of nearly 4.80 dB. Therefore, LDPC code is more effective in improving the link margin as compared to transmit diversity.

It may be mentioned that results presented in Table 7.5 are for a zenith angle of 0°. For other zenith angles, the improvement in power margin with transmit diversity will change. The variations in the link power margin with zenith angle for transmit diversity at $\lambda = 1064$ and 1550 nm are shown in Fig. 7.1a, b, respectively. Using

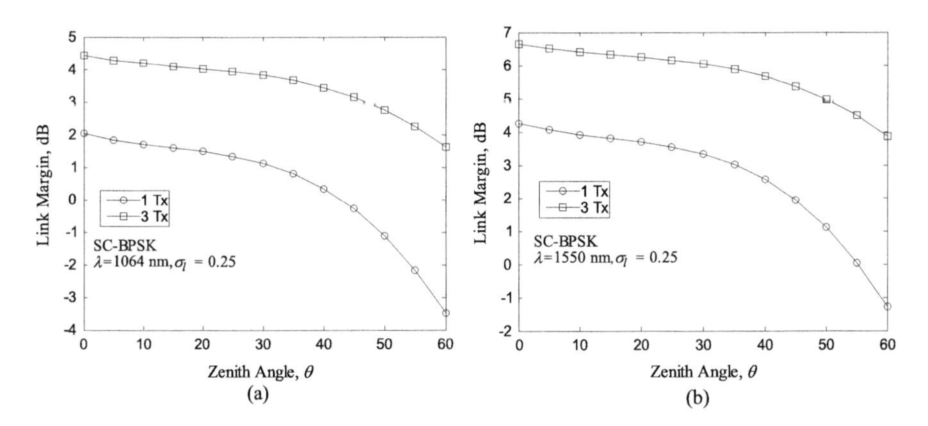

Fig. 7.1 Variations of link margin with zenith angle for SC-BPSK modulation scheme with and without diversity at wavelengths (**a**) $\lambda = 1064$ nm and (**b**) $\lambda = 1550$ nm, respectively

these figures, it is seen that variations of link margin with zenith angle at $\lambda = 1064$ nm and $\lambda = 1550$ nm follow almost a similar trend. At $\lambda = 1064$ nm, the link margin decreases continuously with zenith angle. This is true in case of both single transmitter and transmit diversity ($M = 3$). In case of single transmitter, the link margin is almost constant up to $\theta = 30°$. With the increase in the value of θ above 30°, there is a decrease in the link margin. With transmit diversity, there is not much decrease in the link margin with the increase in the zenith angle. As an example, at a zenith angle $\theta = 30°$, the power margin from Fig. 7.1a will reduce to 1.11 dB from 2.03 dB at zero zenith angle (row 13 and column 5 of Table 7.5). In case of transmit diversity ($M = 3$), the power margin at zenith angle $\theta = 30°$ from Fig. 7.1a will reduce to 3.82 dB from 4.43 dB at zero zenith angle (row 14 and column 5 of Table 7.5). At this zenith angle, usage of LDPC will become more advantageous and provide a gain of 5.45 dB (9.27 − 3.82 dB) instead of 4.80 dB over the transmit diversity ($M = 3$).

7.3 Summary

The performance of FSO communication system depends upon available bandwidth, receiver sensitivity, transmitter and receiver antenna gains, and environmental conditions. Free space loss is one of the dominant component for decrease in the strength of the signal. In order to study the availability of the FSO system, link budget analysis is carried out. This chapter evaluates FSO link budget for SC-BPSK modulation scheme at two operating wavelengths, i.e., $\lambda = 1064$ and 1550 nm. The power budget has been calculated for a link between Earth and geosynchronous orbit satellite, i.e., approx. distance of 40,000 km. Link performance improvement using diversity and error correcting codes is also evaluated. It is seen that link margin using single transmitter is not adequate enough for reliable FSO communication link. With transmit diversity ($M = 3$), an improvement of nearly 2.40 dB in the power margin has been observed at both the wavelengths. Instead of transmit diversity, if LDPC code is used, the corresponding improvement in the power margin is nearly 7.24 dB. Further, the link power margins depend upon the zenith angle. Therefore, variations of power margin with zenith angle ($\theta = 0°$ to $60°$) have been given. It is seen that power margins decrease continuously with zenith angles in both the cases. However, the decrease in power margin with zenith angle is relatively less in transmit diversity as compared to single transmitter. Consequently, the improvement obtained with the transmit diversity becomes more which is 2.71 dB (3.82 − 1.11 dB) at zenith angle $\theta = 30°$ and $\lambda = 1064$ nm. It implies that transmit diversity provides more gain at higher zenith angle, i.e., in adverse conditions. At the same zenith angle and wavelength, if LDPC code is used, it provides the same gain, i.e., 9.27 dB. In that case, the gain provided by the LDPC code over the single transmitter is 8.16 dB (9.27 − 1.11 dB). At both the zenith angles ($\theta = 0°$ and $30°$), the gain provided by the LDPC code is much more than the gain provided by the transmit diversity. It will enhance further with the increase in zenith angle. This is true for both operating wavelengths of 1064 and 1550 nm. It is therefore concluded that LDPC code is more effective in improving the power margin as compared to transmit diversity ($M = 3$).

Bibliography

1. B.J. Klein, J.J. Degnan, Optical antenna gain I: transmitting antennas. Appl. Opt. **13**(9), 2134–2141 (1974)
2. V.A. Vilnrotter, The effect of pointing errors on the performance of optical communications systems, TDA progress report 42–63, Jet Propulsion Laboratory, Pasadena, 1981, pp. 136–146
3. P.W. Gorham, D.J. Rochblatt, Effect of antenna-pointing errors on phase stability and interferometric delay, TDA progress report 42–132, Jet Propulsion Laboratory, Pasadena, Feb 1998, pp. 1–19
4. W.K. Marshall, B.D. Burk, Received optical power calculations for optical communications link performance analysis, TDA progress report 42–87, Communication Systems Research Section, Jet Propulsion Laboratory, Pasadena, Sept 1986

Index

H. Kaushal et al., *Free Space Optical Communication*, Optical Networks,
DOI 10.1007/978-81-322-3691-7

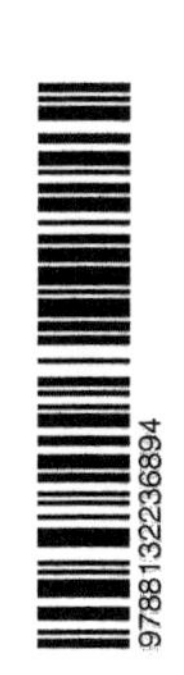
9788132236894